Synthesis and Characterization of Luminescent Cu(I) Complexes

Zur Erlangung des akademischen Grades eines

DOKTORS DER NATURWISSENSCHAFTEN

(Dr. rer. nat.)

von der KIT-Fakultät für Chemie und Biowissenschaften

des Karlsruher Instituts für Technologie (KIT)

genehmigte

DISSERTATION

von

Diplom-Chemikerin Manuela Wallesch

aus

Landshut

KIT-Dekan:	Prof. Dr. Willem M. Klopper
Referent:	Prof. Dr. Stefan Bräse
Korreferent:	Prof. Dr. Clemens Heske
Tag der mündlichen Prüfung:	18.12.2015

Band 58
Beiträge zur organischen Synthese
Hrsg.: Stefan Bräse

Prof. Dr. Stefan Bräse
Institut für Organische Chemie
Karlsruher Institut für Technologie (KIT)
Fritz-Haber-Weg 6
D-76131 Karlsruhe

Bibliografische Information der Deutschen Bibliothek

Die Deutsche Nationalbibliothek verzeichnet diese Publikation in der
Deutschen Nationalbibliografie; detaillierte bibliografische Daten sind
im Internet über http://dnb.d-nb.de abrufbar.

ISBN 978-3-8325-4230-6
ISSN 1862-5681

Logos Verlag Berlin GmbH
Comeniushof, Gubener Str. 47,
10243 Berlin
Tel.: +49 030 42 85 10 90
Fax: +49 030 42 85 10 92
INTERNET: http://www.logos-verlag.de

Die vorliegende Arbeit wurde vom April 2012 bis November 2015 am Karlsruher Institut für Technologie (Institut für Organische Chemie) unter Betreuung von Prof. Dr. Stefan Bräse und Prof. Dr. Clemens Heske durchgeführt.

Table of contents

1 Abstract

Luminescent Cu(I) NHetPHOS complexes are very efficient emitters, which have successfully been deployed in organic light-emitting diodes (OLEDs). They exhibit several advantages in comparison to other Cu(I) complexes: Apart from their high photoluminescence quantum yield, they cover the whole visible spectrum. Furthermore, they offer the possibility to be processed from solution due to good solubility, which allows for cost- and energy-efficient processing techniques such as printing rather than state-of-the-art vacuum processing.

In this work, various important questions concerning Cu(I) complexes were addressed. By means of hard x-ray spectroscopy at the Cu K edge, it was shown that significant photophysical differences, between solution, bulk solid, and thin film samples, are not a result of dissociation and that the studied complexes retain their principle structure even when preparing thin films from solution. Therefore, it was demonstrated that NHetPHOS emitters are suitable to be deployed in solution-processed optoelectronic devices. By means of soft x-ray spectroscopy at the N K edge, ultraviolet photoelectron spectroscopy, and inverse photoemission spectroscopy, the electronic structure with particular focus on the occupied and unoccupied molecular orbitals, formed upon binding of the Cu- and N-atoms, and the HOMO and LUMO energies of these materials were investigated. Furthermore, new red-emitting Cu(I) complexes were developed.

Consequently, a breakthrough on the field of organic light-emitting diodes was enabled: The principle structure of a not crystallizable NHetPHOS complex with a bridging bisphosphine ligand, yielding a new quantum efficiency record for both solution- and vacuum-processed organic light emitting diodes with Cu(I) complexes as emitters, was determined by means of x-ray spectroscopy at the Cu K edge.

Kurzzusammenfassung

Lumineszierende Cu(I)-NHetPHOS-Komplexe sind sehr effiziente Emitter, die bereits erfolgreich in organischen Leuchtdioden (OLEDs) eingesetzt wurden. Cu(I)-NHetPHOS-Komplexe besitzen mehrere Vorteile im Vergleich zu anderen Kupferkomplexen: Sie weisen eine hohe Photolumineszenz-Quantenausbeute auf und decken das gesamt Farbspektrum ab. Darüber hinaus eignen sich diese Materialien durch die gute Löslichkeit dafür, aus Lösung prozessiert zu werden. Dies ermöglicht den Einsatz von kosten- und energieeffizienten Prozessierungsverfahren wie Drucktechniken im Gegensatz zu aufwändiger Vakuum-Prozessierung.

In dieser Arbeit wurden wichtige Fragestellungen bezüglich Cu(I)-Komplexen adressiert. Mittels Harträntgenspektroskopie an der Cu-K-Kante wurde gezeigt, dass deutliche Unterschiede in den photophysikalischen Eigenschaften von Lösungs-, Feststoff- und Filmproben nicht auf Dissoziation zurückzuführen sind und die untersuchten Komplexe bei Flüssig-Prozessierung ihre Grundstruktur beibehalten. Damit wurde dargelegt, dass NHetPHOS-Komplexe für die Prozessierung aus Lösung geeignet sind. Mittels Weichröntgenspektroskopie an der N-K-Kante, Ultraviolett-Photoelektronenspektroskopie und inverser Photoemissionsspektroskopie wurde die elektronische Struktur mit Fokus auf die besetzten und unbesetzten Molekülorbitale, die im Zuge der Koordination zwischen den Cu- und N-Atomen gebildet werden, und die HOMO- und LUMO-Energien dieser Materialien untersucht. Zudem wurden neue rot-emittierende Cu(I)-Komplexe entwickelt.

Damit konnte ein Durchbruch auf dem Gebiet der organischen Leuchtdioden ermöglicht werden: Die Grundstruktur eines nicht-kristallisierbaren NHetPHOS-Komplexes mit überbrückendem Bisphosphin-Ligand, der einen neuen Quanteneffizienzrekord für sowohl flüssig- als auch vakuumprozessierte OLEDs mit Cu(I)-Komplexen erzielte, wurde mittels Harträntgenspektroskopie an der Cu-K-Kante aufgeklärt.

2   Introduction to luminescent Cu(I) complexes in organic light emitting diodes

2.1   Organic light emitting diodes

2.1.1 Application

According to a market forecast from IDTechEx, sales revenue for plastic and flexible displays based on OLEDs have the potential to grow from about $ 2 bn in 2015 to about $ 16 bn in 2020.[1] Automotive and aerospace, TV sets, and wearable electronics are the market segments that are expected to exhibit the highest growth rates for OLED displays.[1] The potential increase of market share for OLEDs in proportion to liquid crystal displays (LCD), which are currently widely-used in consumer electronics, is due to a number of advantages of OLEDs, which originate in the setup of the device.

Organic light emitting diodes (OLEDs) are light emitting diodes, which can transform electrical energy into visible light and in which the emissive layer (EL) is composed of an organic or metal organic material. A schematic depiction of an idealized OLED architecture with a transparent substrate, a transparent anode, and a reflective cathode is given in Figure 1. In-between the electrodes are the emissive layer (EL), and additional organic layers, e.g., charge injection, transport, and blocking layers. Each layer has a thickness of 10 to 100 nm. Indium tin oxide (ITO) on a glass or polymer substrate is often used as anode material, as it is transparent and conductive. The organic layers on top of the ITO can either be processed from solution or by vapor deposition. While the fabrication of OLEDs from solution is predicted to be of great relevance in the future, it has not been realized in commercial devices yet and has only been used for R&D test devices so far. The metal cathode is commonly vapor-deposited. To shield the OLED from moisture and oxygen the device has to be encapsulated.

LC displays often consist of a layer of liquid crystals in front of a LED light source. As there is no backlight required for OLED displays, these displays are thinner and have a higher contrast than LC displays. Furthermore, OLED displays can be flexible in contrast to LCD displays because the former are based on non-crystalline films of organic and metal organic emitters and the latter on crystalline inorganic semiconductors. This opens new market segments for OLEDs, which cannot be accessed with LC displays. Major drawbacks for OLED displays are shortcomings concerning the long-term reliability, limitation of the maximum display size, and high production costs, as commercially available OLEDs are processed by vapor deposition.

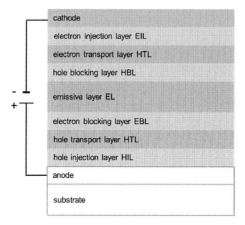

Figure 1: Schematic depiction of an idealized OLED setup with a transpartent substrate, a transparent anode, and a reflective cathode. In-between the electrodes are the emissive layer (EL) and additional organic layers.

By developing solution processed OLEDs to readiness for marketing, the aforementioned drawbacks could be overcome and OLED technology could realize the forecasted market growth.

2.1.2 Principle

The function principle of an OLED is depicted in Figure 2. The conversion of electrical energy into visible light in an OLED can be divided into four steps:

(1.) Charge carrier injection. Holes and electrons are injected from the electrodes into the injection layers.

(2.) Charge carrier transport. Holes and electrons are transported through the transport and blocking layers into the emissive layer.

(3.) Formation of excitons. Due to the blocking layers, the charge carriers accumulate in the emissive layer and form excitons.

(4.) Recombination of holes and electrons. Within the emissive layer, excitons recombine under emission of photons. To obtain an efficient OLED, the HOMO and LUMO energies of neighboring layers have to match. Below, the function principle of OLEDs will be discussed in more detail.

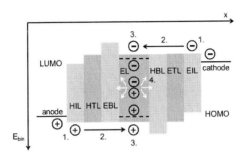

Figure 2: Schematic depiction of the function principle of an idealized OLED. Energy level alignment at the interfaces is not depicted for clarity. 1. Charge carrier injection. Holes/electrons are injected from the electrodes into the HOMO levels of the hole injection layer (HIL)/LUMO levels of the electron injection layer (EIL). 2. Charge carrier transport. Holes and electrons are transported through the hole transport (HTL) and electron blocking layer (EBL) as well as the electron transport (ETL) and the hole blocking layer (HBL). 3. Formation of excitons. Due to the blocking layers the charge carriers accumulate in the emissive layer (EL) and form excitons. 4. Recombination of holes and electrons. Within the EL, excitons recombine under emission of photons.

An external voltage, typically in the range of $2 - 10$ eV, is applied to inject charge carriers from the electrodes into the organic layers (step (1.)). The required voltage for charge carrier injection is determined by the energy of the emitted photons, the thermal relaxation energy of excited states, potential drops due to charge transport, and barriers at the interfaces of the different layers. In general, charge carrier injection barriers are observed for multilayers of organic semiconductors on metals. Different models were used to describe this phenomenon.[2-16] Figure 3 depicts the energy alignment between the Fermi energy of a metal electrode and the HOMO and LUMO energy levels of an organic semiconductor according to the model of Oehzelt et al.[17] This electrostatic model predicts that the density of states (DOS) of the organic semiconductor is the decisive factor controlling the height of the injection barrier.[17] The left part in Figure 3 presents an example for an initial situation of an organic semiconductor on a metal surface. The ionization potential (IP) and the electron affinity (EA) of the organic semiconductor are defined by the energy differences between the onsets of the DOS of the HOMO/LUMO and the vacuum level, respectively. The work function of a sample surface is given by the difference between the Fermi level and the vacuum level. In this case, initially, the work function of the film surface is higher than the IP of the organic semiconductor. Before charge equilibration, electrons reside at energies above the Fermi level. The right part in Figure 3 presents the final situation after the establishment of a common electronic potential and the exchange of charge across the interface. When no external voltage is applied, the Fermi level is constant throughout the sample, since otherwise a current would flow. Upon charge equilibration, charges build up on both sides of the interface, which results in an electric field and a potential gradient. With an increasing number of layers on the metal surface, the DOS shifts to higher binding energies. Hence, with increasing distance from the metal surface, the charge injection barrier Δ_h for holes increases and the injection barrier Δ_e for electrons decreases.[17] It can be assumed that the situation for metal organic films of, e.g., Cu(I) complexes on metal surfaces is similar to that of organic semiconductors.

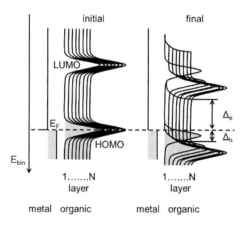

Figure 3: Energy level alignment at interfaces between a metal surface and an organic film.[17]

The charge carrier injection can be described as hopping process from the metal Fermi level into localized states of the organic material according to the model of Scott and Malliaras.[18, 19]

In general, the transport mechanisms in organic amorphous solids are not yet well understood. However, there are different models describing charge charrier transport (step

(2.)) as hopping process in analogy to charge carrier injection[20]. The model first described by Bässler[19] is widely-used. It assumes hopping of charge carriers in an amorphous solid, in which the transport sites energies and separations are in Gaussian distribution. Hopping is field assisted and thermally activated within the ensemble. Therefore, the charge carrier mobility depends on the electric field and the temperature.[19] For organic and metal organic materials, the charge carrier mobility is in the order of 10^{-5} to 10^{-1} $cm^2V^{-1}s^{-1}$.[21]

Driven by the external field, holes and electrons accumulate in the emissive layer, where they form excitons (3.). This process is due to the Coulomb interaction between holes and electrons. According to the Langevin theory,[22] a hole and electron that approach each other within a distance, such that their Coulomb binding energy exceeds k_BT, will ultimately recombine. This distance is known as Onsager radius.

$$r = \frac{e^2}{4\pi\epsilon_0\epsilon_r k_B T} \tag{2.1}$$

e is the elementary charge, ϵ_0 is the vacuum permittivity, and ϵ_r is the dielectric constant. For organic materials with $\epsilon_r = 3$ the Onsager radius is about 19 nm at room temperature.[23]

The spin of the electron is $s = 1/2$. Considering holes as missing electrons, holes are particles with the same properties as electrons, except that they carry a positive charge. Therefore, the spin of holes is $s = 1/2$. In a quantum mechanical treatment, the spins of a hole and an electron can be coupled to four new combined states: One singlet state and three triplet states $s = 0, 1$. It can be assumed that the exciton formation is spin-independent as implicit in the Langevin theory.[24] According to statistics, all excitonic spin states will be equally populated. Consequently, excitons are formed with spin wave functions in the triplet and singlet configuration in the ratio 3:1.

Upon decay of the exciton or recombination of the hole and the electron, energy is released through either radiative or non-radiative pathways. For most organic molecules, the decay of singlet excitons via fluorescent emission from the first excited singlet state to the singlet ground state ($S_1{\rightarrow}S_0$) is an efficient process, whereas phosphorescent emission from the first excited triplet state to the singlet ground state is forbidden ($T_1{\rightarrow}S_0$) and the $S_1{\rightarrow}T_1$ intersystem crossing rate is small. Therefore, in this case the deactivation of the T_1 state occurs via non-radiative paths at ambient temperature and 75% of the excitons are lost. Several strategies to harvest triplet excitons and achieve higher efficiencies in OLEDs have been developed. Firstly, triplet-triplet annihilation describes the Dexter energy transfer between two triplet excited-state molecules, yielding one singlet excited-state molecule and one molecule in the ground state. This is possible, as in general, the energy gap between S_0 and T_1 is bigger than the energy gap between T_1 and S_1. After the annihilation process, the energy level that the electron occupies will be twice the energy difference between the first excited triplet state and the singlet ground state $\Delta E(T_1.S_0)$. After non-radiative decay, fluorescent emission from the S_1 can occur. The maximum quantum efficiency in this case is 62.5%. Secondly, heavy metals with high spin-orbit coupling in the emissive material facilitate intersystem crossing and the emission from the first excited triplet state to the singlet ground state $T_1 \rightarrow S_0$ occurs readily.[25] This concept is referred to as triplet harvesting. In this case, the maximum quantum efficiency is 100%. Thirdly, singlet harvesting can occur when the energy difference $\Delta E(S_1\text{-}T_1)$ between the first excited singlet state S_1 and the first excited triplet state T_1 is small and in the range of $k_B T$. At room temperature, the S_1 state can be repopulated from the T_1 efficiently and emission from the first excited singlet state S_1 can occur with a maximum quantum efficiency of 100%.[25] This process is also called thermally activated delayed fluorescence (TADF).

2.2   Luminescent Cu(I) complexes[*]

2.2.1 Structural motifs in Cu(I) complexes

Cu(I) complexes are highly efficient TADF emitters and therefore have been studied as emitters in OLEDs.[27-31]

In general, mononuclear Cu(I) complexes feature a tetrahedral or pseudo-tetrahedral coordination environment. "The core of multinuclear Cu(I) halide complexes often consists of rhomboid subunits with Cu(I) and halide atoms at alternate corners. In multinuclear Cu(I) complexes, rhomboid subunits are connected or fused by sharing corners or edges (Figure 4).[32] The chemistry of most Cu(I) complexes is dominated by the strongly

[*] Parts of this chapter were previously published in the context of this work: [26]   M.   Wallesch,   D. Volz, D. M. Zink, U. Schepers, M. Nieger, T. Baumann, S. Bräse, *Chem. Eur. J.* **2014**, *20*, 6578-6590.

favored, tetrahedral or pseudo-tetrahedral coordination geometry. In this situation, each Cu(I) ion has two open coordination sites in a dinuclear complex. When bulky ligands are used, the Cu(I) ions prefer trigonal-planar coordination geometry, affording $[L_3Cu_2X_2]$ or $[L_2Cu_2X_2]$ structures.

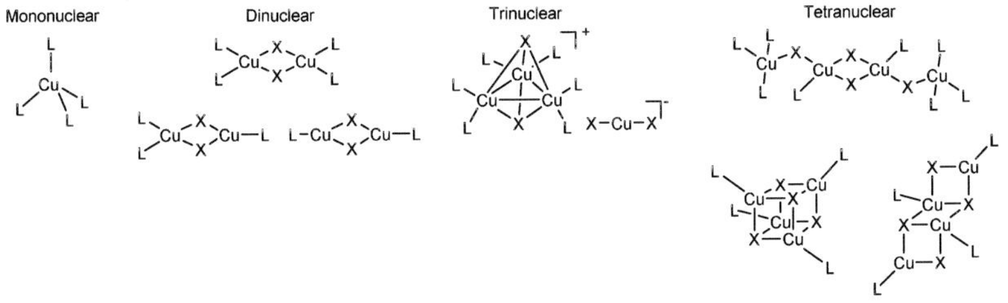

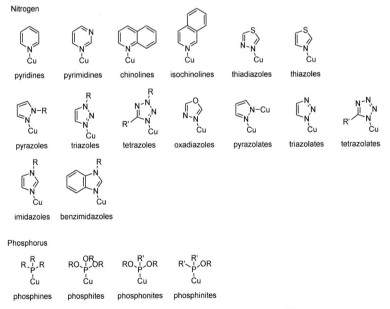

Figure 4: "Common structures of multinuclear Cu(I) complexes.[32],[26]

Trinuclear complexes often consist of a positively charged $[Cu_3X_2]^+$ ion, which arrays in a trigonal bipyramid and a linear $[CuX_2]^-$ counterion. For tetranuclear complexes, two main forms are possible: heterocubane-structures or open, step-like structures (Figure 4)."[26]

Figure 5: Selected motifs found in N and P ligands for Cu(I) complexes.[26]

The synthesis of Cu(I) complexes is often carried out at room temperature by a simple mixing of a suitable Cu precursor and a ligand. "In many cases, the turn-over is nearly quantitative and high yields may be achieved. Cu(I) complexes with a broad variety of coordinating atoms like N, O, P, S, Se, C, and As have been reported. As predicted by Pearson's HSAB-concept,[33] the soft Cu(I) ion is better stabilized upon coordination with soft atoms, which has been demonstrated by Samuelson et al. for Cu(I) complexes with phosphine and phosphine oxide ligands.[34] Nitrogen and phosphorus containing ligands are favorable, as they are soft electron donors that are accessible from commercially available precursors"[26] (Figure 5).

2.2.2 Cu(I) complexes as emitters in OLEDs

Cu(I) complexes have to meet certain requirements for the application as emitters in OLEDs, which are a high photoluminescence quantum yield, indicating a potentially high electroluminescence quantum yield, as well as chemical stability under processing conditions.

"In general, the photoluminescence quantum yield for Cu(I) halide complexes with only phosphine ligands is low."[26] Complexes of Cu(I) halides with pyridine derivatives and PPh_3 (Figure 6) have been studied intensively due to their good photoluminescence quantum yields.[35-42]

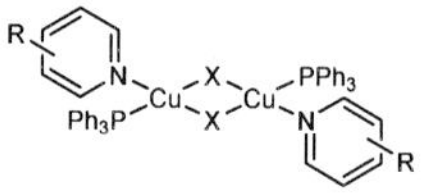

X = Cl, Br, I, CN

R = e.g. H, Me, vinyl, CN, RC=O ...

Figure 6: "Heteroleptic Cu(I) complexes with pyridine and phosphine ligands have been studied for four decades. Various halides and pseudo halides as well as substituents are tolerated."[26]

However, these complexes "have not been published as emitters in OLEDs until today. The main reason for this is their intrinsic lability: The photophysical properties often change drastically upon processing. Upon drying under reduced pressure (which is necessary when preparing thin films for OLEDs), volatile ligands such as pyridine may be extruded, while species such as $[CuX(PR_3)]_n$ are formed.[43],[26]

Bis(diphenylphosphino)isopropyl amine (dppipa) was introduced as ligand for Cu(I) complexes by Samuelson and coworkers.[44] Due to its steric flexibility it can serve as a chelating as well as a bridging ligand. "Due to this, mono-, di-, tri-, and tetranuclear Cu(I) dppipa complexes have been reported (see Figure 7)."[26] The different forms could be obtained selectively, controlled by the stoichiometry.[44] However, dppipa is sensitive towards oxidation and hydrolysis.[26] For the application in OLEDs it has to be ensured that the Cu(I) complexes are not oxidized and hydrolyzed upon processing.

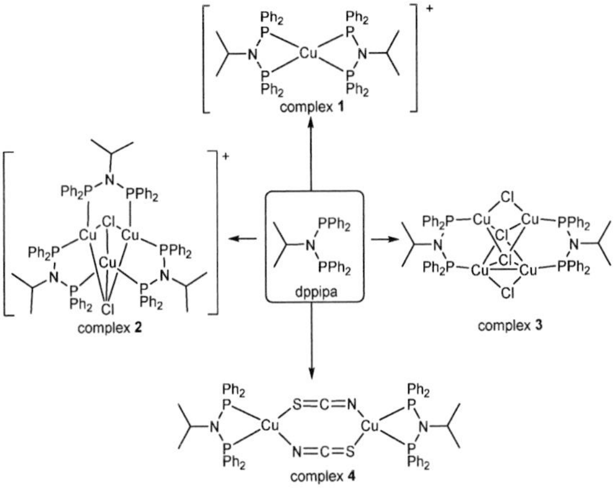

Figure 7: "Samuelson and coworkers introduced dppipa as ligand. Depending on the stoichiometry, mono- to tetranuclear complexes were obtained.[44],[26]

"In general, Cu(I) complexes tend to reversibly dissociate in solution, which is one of the key properties that led to the application of Cu(I) complexes as catalysts in organic synthesis."[26] If more than one ligand is deployed, besides the desired heteroleptic complexes homoleptic complexes may be formed in solution.[26] Upon quick precipitation, such mixtures might form species that differ from the ones obtained by slow crystallization.[26]

The consequences are important when it comes to the application as emitters in OLEDs: The photophysical properties of Cu(I) complexes can change when moving from bulk samples to thin films.[26] "Nierengarten and coworkers recently showed that complexes of the type [Cu(N,N)(P,P)]+ dissociate in solution.[45] In this particular case, Cu(I) bisphenanthroline complexes are formed, which are actually highly absorbent and currently under investigation as absorber materials for dye-sensitized solar cells.[46, 47] Even small traces of potential Cu(I) bisimine byproduct are likely to reduce the performance of a light emitting device."[26]

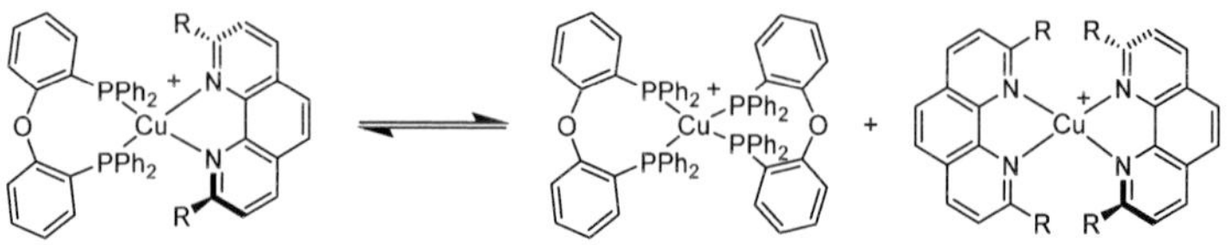

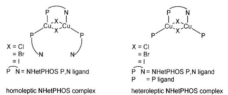

complex **5**
desired product
high quantum efficiency

complex **6**
degradation product 1
low quantum efficiency

complex **7**
degradation product 2
potent quencher

Figure 8: "Nierengarten and coworkers showed that complexes of the type [Cu(N,N)(P,P)]⁺ dissociate in solution, which leads to the formation of highly absorbant dyes of the type [Cu(N,N)₂]⁺, depending on the size of the substituent R. Even when the formation of degradation products 1 and 2 is minimized by choosing proper P,P and N,N ligands, the device efficiency is potentially hindered due to the formation of highly absorbant Cu(I) bisimine complexes.[45]".[26]

Recently, a family of dinuclear Cu(I) halide complexes with NHetPHOS-type ligands has been developed in the Bräse group.[28, 29, 48] These complexes consist of a butterfly-shaped Cu_2X_2 core, one bidentate NHetPHOS ligand, which is bridging the two Cu(I) centers, and two terminal ligands. The terminal ligands can either be NHetPHOS ligands, which only coordinate via their phosphorus atom or other monodentate P ligands. Each Cu(I) atom features a pseudo-tetrahedral coordination environment. Lacking more precise nomenclature, complexes in which the bridging ligand and the terminal ligands are the same NHetPHOS ligand are referred to as homoleptic, despite the fact that the complexes also contain halide ligands. In the same way, complexes in which the bridging ligand and the terminal ligands are different are called heteroleptic (see Figure 9).

X = Cl
= Br
= I
P⌒N = NHetPHOS P,N ligand

homoleptic NHetPHOS complex

X = Cl
= Br
= I
P⌒N = NHetPHOS P,N ligand
P   = P ligand

heteroleptic NHetPHOS complex

Figure 9: Generalized structures of homoleptic and heteroleptic NHetPHOS Cu(I) complexes.[28, 29, 48]

Furthermore, instead of pyridine, a variety of other N-heterocycles, e.g., quinoline, isoquinoline, triazoles, benzimidazoles, thiazoles, or benzothiazoles were used in N,P ligands.[28, 48, 49] Exemplarily, NHetPHOS complexes **1–6** are depicted in Figure 10.

NHetPHOS complexes cover the whole visible spectrum but there are only few complexes with an efficient red emission. For this reason, red-emitting Cu(I) complexes will be developed in this work.

NHetPHOS complexes were deployed as emitter in OLEDs due to their high photoluminescence quantum yield. The solubility of NHetPHOS complexes in appropriate solvents can be advantageous, as processing of OLEDs from solution is more cost-effective than by vapor deposition. Whereas the efficiency of devices with solution-processed ELs containing Cu(I) NHetPHOS complexes is high, the device lifetime has to be improved for commercial applications. Therefore, it is important to obtain a better overall understanding of these materials. The molecular structure of Cu(I) NHetPHOS complexes in non-crystalline films as well as the electronic structure of these materials is studied herein.

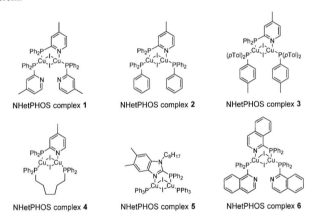

Figure 10: Molecular structures of NHetPHOS complexes **1–6**.

The possibility of Cu(I) complexes to undergo chemical reactions upon solution-processing is studied in this work. Accordingly, different methods for the characterization of the molecular structure of Cu(I) NHetPHOS complexes and their relevance for the processed material in an OLED are evaluated. Suitable techniques are chosen to study the molecular structure of Cu(I) NHetPHOS complexes in non-crystalline form.

In general, the luminescence of Cu(I) complexes can be attributed to d-s transitions on the metal centers, ligand-centered transitions, inter-ligand transitions, as well as charge transfer transitions between the metal center and ligands. As the emission color of Cu(I) NHetPHOS complexes can be tuned by systematical variation of the ligand system, it was assumed that metal-to-ligand charge-transfer (MLCT), ligand-to-ligand charge-transfer (LLCT) and/or metal-halide-to-ligand charge-transfer (M+X)LCT transitions are involved in the emission process.[48, 50, 51] However, the electronic structure of NHetPHOS complexes has not been studied experimentally in detail so far. A greater insight into the electronic interaction between the Cu(I) atom and the ligands is required to further improve the design of these materials and thereafter yield solution-processed OLEDs with longer device lifetimes.

Furthermore, the HOMO and LUMO energies of the Cu(I) emitter are important parameters for OLED fabrication, as they have to match the HOMO and LUMO energies of the neighboring layers (see 2.1.2). As there are different methods to assess these parameters, the significance of these methods for solution-processed Cu(I) NHetPHOS emitters is reviewed herein, on the basis of a study of chosen members of the Cu(I) NHetPHOS complex family.

3   Aims and objectives

Cu(I) NHetPHOS emitters exhibit high photoluminescence quantum yields and cover the whole visible spectrum. Whereas there is a multiplicity of green- and yellow-emitting representatives, only few examples for red emitters have been reported. Cu(I) NHetPHOS complexes are efficient emitters in OLEDs with emissive layers processed from solution. However, the OLED device performance needs to be improved for commercial applications. To achieve this, a better overall understanding of the Cu(I) NHetPHOS complexes is required.

The aims of this work are to study the molecular structure of Cu(I) NHetPHOS complexes in non-crystalline form and upon solution-processing and obtain insight on the electronic structure with particular focus on the HOMO and LUMO energies of these materials and new develop new red-emitting Cu(I) complexes. The scope of this work is divided into four objectives.

- For the development of red-emitting Cu(I) complexes, appropriate ligands are identified and synthesized. These ligands are reacted with Cu(I) salts to yield new Cu(I) complexes. The chemical and photophysical properties of these materials are characterized.
- Different methods for the characterization of the molecular structure of Cu(I) complexes – in particular, single crystal x-ray diffraction, photophysical measurements, IR and NMR spectroscopy, and x-ray absorption spectroscopy at the Cu K edge – are evaluated. Based on this evaluation, appropriate methods are chosen to study selected Cu(I) NHetPHOS complexes. In the context of this work, x-ray absorption spectroscopy at the Cu K edge is used to characterize non-crystalline samples of Cu(I) NHetPHOS complexes and investigate whether these materials are stable upon film processing or undergo chemical reactions as reported for other Cu(I) complexes.
- The electronic structure of Cu(I) NHetPHOS complexes is investigated by means of resonant inelastic soft x-ray scattering and x-ray emission spectroscopy at the N K edge. The emphasis of this study will be directed towards the occupied and unoccupied molecular orbitals formed upon coordination of the Cu and N atom, which are expected to influence the luminescence properties of the Cu(I) NHetPHOS complexes.

- The HOMO and LUMO energies of the Cu(I) emitter are important parameters for OLED fabrication, i.e., the construction of a durable, efficient stack architecture and can be assessed by different methods. The significance as well as advantages and limitations of different methods for the evaluation of the HOMO and LUMO energies of Cu(I) NHetPHOS complexes – in particular optical spectroscopy, cyclic voltammetry, photoelectron spectroscopy in air, and photoelectron spectroscopy –will be reviewed on the basis of a study of selected members of the Cu(I) NHetPHOS complex family.

4 Main section: Synthesis and characterization of Cu(I) NHetPHOS complexes

4.1 Synthesis and characterization of red-emitting Cu(I) complexes

4.1.1 Cu (I) complexes with quinoline and (Ph)₂POP(Ph)₂ ligands

Symmetric, dinuclear heteroleptic complexes of the type NPCuXXCuNP (see Figure 11) were first structurally investigated by the group of White and coworkers.[41, 52-57] Recently, Tsuge et al. reported that these materials are also interesting emitting materials and that their color can be tuned by varying either the halides or the N-heterocycles.[58] From quantum-chemical calculations, it is known that the emissive transition for similar complexes is a result of a metal-halide-to-ligand charge transfer (M+XLCT) between the dinuclear copper halide subunit and the N-heterocycles.[42, 59]

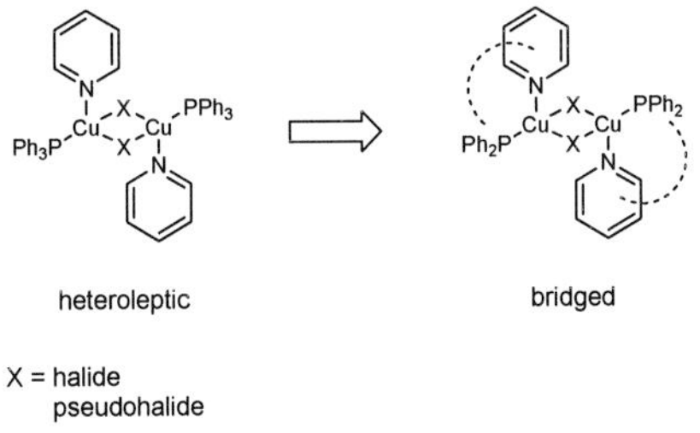

heteroleptic   bridged

X = halide
pseudohalide

Figure 11: Modification of heteroleptic Cu(I) complexes by introduction of molecular bridges. The color may be tuned by varying either the halides X or the N-heterocycles.[58]

The influence of the phosphine on the emission properties is negligible.[28, 42, 59] Because of the current lack of efficient red copper emitters, quinoline was identified as potentially interesting ligand due to its narrow HOMO-LUMO separation, high chemical stability and encouraging previous results with similar structures.[42, 48, 58] Potential stability problems with such heteroleptic complexes, which are the result of a cleavage of the Cu-N bond, were reported previously.[43] The instability of these complexes may hamper their durability in OLED devices. It can be assumed, that the introduction of inert bridges between the phosphine and N-heterocyclic ligands can stabilize those dinuclear structures with kinetically favored five- or six-membered metallacycles.[60] A similar strategy was introduced by Yersin and Wesemann, where direct fusion of amines with phosphines where yielded five-membered rings with Cu(I).[61] Potential bridges are short alkyl chains, ether or thioether bridges. In a first attempt, a single oxygen atom was chosen as bridging unit between diphenylphosphine and quinoline.

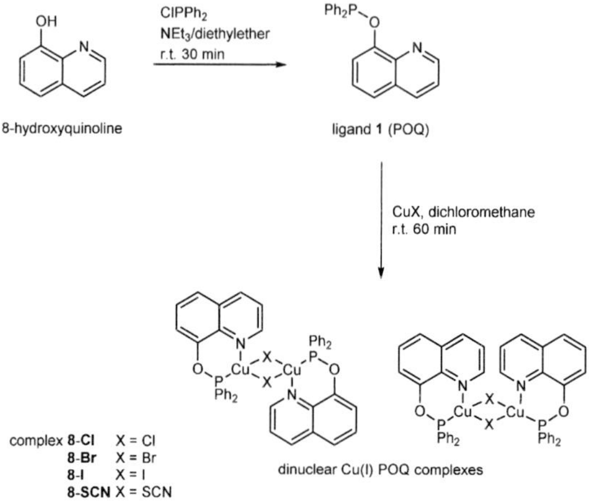

Scheme 1: Synthesis of ligand **1** (POQ) and complexation with copper precursors yields dinuclear complexes.

Fusing phosphine and N-donor moieties in one chelating ligand was achieved by combination of diphenylphosphine as P-donor, quinoline as N-donor, and oxygen as the bridging unit, which yielded ligand **1** (POQ, 8-((diphenylphosphanyl)oxy)-quinoline, see Scheme 1). The synthesis of the ligand is straight-forward and could easily be performed in large batches by reacting equimolar amounts of chlorodiphenylphosphine and 8-hydroxyquinoline in dry trimethylamine. Ligand **1** could be purified by multiple recrystallizations to get a pure product. The synthesis of the Cu(I) complexes was performed in dry dichloromethane by mixing of stoichiometric amounts of Cu(I) precursors with ligand **1** and precipitation of the product. For the synthesis of Cu(I) complexes, non-crystallized ligand **1**, which can be obtained by filtering the reaction mixture over celite and removal of the solvent, was deployed.

Reaction of POQ with precursor salts CuX (X =Cl, Br, I, SCN) was performed as previously described in the literature.[49] The products were obtained by diffusion of cyclohexane into the clear, yellow reaction mixtures or by precipitation from the reaction mixture with hexane (bulk samples) to yield the complexes **8–Cl**, **8–Br**, **8–I**, and **8–SCN**.

The photophysical properties of complexes **8–Cl**, **8–Br**, **8–I**, and **8–SCN** were studied in solid state under nitrogen to prevent triplet quenching from oxygen. Only for complex **8–Br** luminescence was detected at room temperature. Upon cooling to 77 K (liquid nitrogen bath), all samples showed weak or very weak luminescence. However, only complexes **8–Cl** and **8–Br** gave satisfactory PL spectra (Figure 12). At room temperature, the photoluminescence quantum yield of complex **8–Cl** was below 1%, while complex **8–Br** yielded a mediocre PLQY of 10 ± 1%.

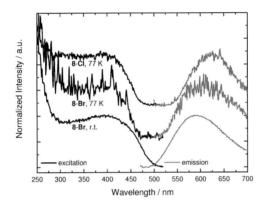

Figure 12: Excitation ($\lambda_{Em.}$ = 600/625 nm) and emission ($\lambda_{Ex.}$ = 370/390 nm) spectra of powder samples of complex **8–Br** at r.t. and 77 K/complex **8–Cl** at 77 K.

All emission spectra are broad and unstructured (see Figure 12), as it is expected for (M+X)LCT emission. The peak in emission intensity of complex **8–Br** is in the orange-red region around 590 nm at room temperature and around 610 nm at 77 K, while the peak in emission intensity for complex **8–Cl** is around 625 nm at 77 K. All spectra have a broadness of 0.4 ± 0.1 eV. Similar to related complexes introduced by Tsuge and coworkers,[42, 59, 62] the large π-system of the quinoline moiety leads to a stabilization of the LUMO and consequently to a red shift, which can be further enhanced by destabilizing the HOMO, by moving from bromide to chloride. Also, based on the fact that the color is red-shifted upon cooling for complex **8–Br** and that the PLQY seems to be increased at lower temperatures, it can be speculated that the material is a TADF emitter as other dinuclear Cu(I) complexes.[61] To confirm this assumption, a deeper spectroscopic analysis is necessary.

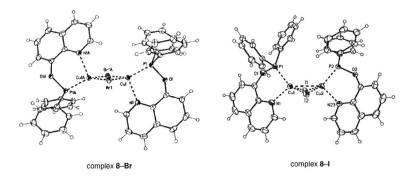

complex **8–Br**   complex **8–I**

Figure 13: Molecular structure of complex **8–Br**. and **8–I**. Displacement parameter drawn at the 50% probability level

For the application of Cu(I) POQ emitters in OLEDs, the PLQY should be higher. The fact that the PLQY is low, but is increased upon cooling to low temperatures for complexes **8–Cl**, **8–Br**, **8–I**, and **8–SCN**, is a hint that strong, temperature-dependent quenching processes present.

Table 1: Selected bond lengths in Å for **8–Br** and **8–I**.

complex	**8–Br**	**8–I**
Cu–X	2.4348 (4)	2.6097(5)
	2.4888 (6)	2.6600(6)
Cu–N	2.132 (2)	2.135(3)
Cu–P	2.1532 (7)	2.1789(9)
P–O	1.6617 (18)	1.652(2)
Cquin–O	1.386 (3)	1.380(4)

The fact that the bromide derivative has the highest PLQY is noticeable. In most cases, the PLQY either increases or decreases when moving from Cl over Br to I.[42, 48, 59] The different behavior in this case might be due to differences in the molecular structure of the samples.

Table 2: Bond distances in Å in various representative Cu(I) compounds from the literature.

compound	ref.	bond distance / Å	compound	ref.	bond distance / Å
Cu-I			Cu-Br		
Cu(I) iodide	[63]	2.6414	Cu(I) bromide	[63]	2.5201
[CuI(pyridine)]$_n$	[64]	2.641(1)–2.689(1)	[CuBr(pyridine)]$_n$	[64]	2.450(5)–2.557(3)
[CuI(PPh$_3$)]$_4$	[65]	2.653(3)–2.732(3)	[CuBr(PPh$_3$)]$_4$	[66]	2.491(2)–2.617(2)
[CuI(pyridine)(PPh$_3$)$_2$]	[67]	2.636(1)	[CuBr(pyridine)(PPh$_3$)$_2$]	[67]	2.459(2)
[CuI(quinoline)]$_2$	[68]	2.657(3)–2.668(2)	[CuBr(quinoline)]$_2$	[69]	2.493(3)–2.680(3)
[CuI(quinoline)]$_4$	[67]	2.680(6)–3.394(12)	[CuBr(quinoline)]$_4$	[70]	2.381(4)–2.851(2)
Cu-Cl			Cu-N		
Cu(I) chloride	[63]	2.3820	[CuI(pyridine)]$_n$	[64]	2.038(6)
[CuCl(pyridine)]$_n$	[64]	2.338(4)–2.530(2)	[CuBr(pyridine)]$_n$	[64]	2.00(1)
[CuCl(PPh$_3$)]$_4$	[71]	2.363(2)–2.505(2)	[CuCl(pyridine)]$_n$	[64]	1.993(8)
[CuCl(pyridine)(PPh$_3$)$_2$]	[67]	2.318(2)	[CuSCN(pyridine)]$_n$	[72]	2.054(5)–2.103(2)
[CuCl(quinoline)]$_2$	[69]	2.357(7)–2.675(6)	[CuI(pyridine)(PPh$_3$)$_2$]	[67]	2.131(5)
[CuCl(quinoline)]$_4$	[70]	2.311(5)–2.510(5)	[CuBr(pyridine)(PPh$_3$)$_2$]	[67]	2.145(9)
Cu-P			[CuCl(pyridine)(PPh$_3$)$_2$]	[67]	2.130(2)
[CuI(PPh$_3$)]$_4$	[65]	2.251(6)–2.258(7)	[CuSCN(pyridine)(PPh$_3$)$_2$]	[73]	2.070(1)–2.091(2)
[CuBr(PPh$_3$)]$_4$	[66]	2.206(3)–2.209(3)	[CuI(quinoline)]$_2$	[68]	2.076(5)–2.104(5)
[CuCl(PPh$_3$)]$_4$	38	2.192(2)–2.193(2)	[CuBr(quinoline)]$_2$	[69]	2.01(1)–2.07(2)
[CuI(pyridine)(PPh$_3$)$_2$]	[67]	2.283(2)–2.292(2)	[CuCl(quinoline)]$_2$	[69]	2.00(2)–2.08(2)
[CuBr(pyridine)(PPh$_3$)$_2$]	[67]	2.277(3)–2.286(4)			
[CuCl(pyridine)(PPh$_3$)$_2$]	[67]	2.256(1)–2.272(1)			
[CuSCN(pyridine)(PPh$_3$)$_2$]	[73]	2.1974(5)			

The molecular structure of complexes **8–Br** and **8–I** could be determined with single crystal x-ray diffraction (Figure 13). As expected, both complexes consist of a dinuclear

Cu$_2$X$_2$ unit with two molecules of POQ chelating the Cu(I) atoms. However, there are fundamental differences in the two structures: Complex **8–Br** is a centrosymmetric complex with a planar Cu$_2$Br$_2$ unit, which is similar to the related complex published by Tsuge et al.[58] On the other hand, complex **8–I** has mirror symmetry and features a bent, butterfly-shaped Cu$_2$I$_2$ unit, much like the NHetPHOS Cu(I) complexes.[29, 60, 74]

Table 3: Bond distances (C-O and P-O) in Å in selected compounds.

compound	ref.	bond distances
quinoline derivatives		
8-hydroxyquinoline	[75]	C-O 1.367(5)
5-chloro-8-hydroxyquinoline	[75]	C-O 1.346(7)
POQ complexes		
[Fe(POQ)(CO)$_2$Br$_2$]	[76]	P-O 1.625, C-O 1.387
[Ni(POQ)$_2$]$^{2+}$	[77]	P-O 1.632, C-O 1.382
POP complexes		
[(CO)$_5$Mo(μ-POP)Mo(CO)$_5$]	[78]	P-O 1.6410(15)
[Mo(CO)$_5$(POP)Fe(CO)$_4$]	[79]	P-O 1.641(3), 1.630(3)
[Mo(CO)$_3$I$_2$(POP)]	[80]	P-O 1.668(3), 1.638(3)
[Mo(CO)$_3$Br$_2$(tolyl$_2$POPtolyl$_2$)]	[80]	P-O 1.59(3), 1.62(3)
[(Cu$_4$(μ$_3$-Cl)$_2$(μ$_2$-Cl)$_2$(μ$_2$-(POP)$_2$]	[81]	P-O 1.646(4) 1.648(3)
[Cu$_2$(μ$_2$-Cl)$_2$(μ$_2$-POP)(η$_1$-PPO)(PPh$_3$)]	[81]	P-O 1.644(3), 1.651(3)

A similar behavior was found for other dinuclear Cu(I) complexes by Yersin and Wesemann, where a bridging N,P ligand yielded a centrosymmetric complex, similar to **8–Br** with chloride as halide, and a mirror symmetric complex, similar to **8–Br** with bromide and iodide as halide.[61] This could be a reason for the abnormal trends in the PLQY when moving from Cl to Br to I.

Even though fusing of diphenylphosphine and quinoline is feasible and yielded orange-red-emitting complexes with Cu(I) halides, there are issues concerning the general applicability of complexes with fused P- and N-donor moieties as emitters: First, as described above, the PLQY should be higher. Second, there are abnormalities in the bond lengths, which might open up degradation pathways for Cu(I) complexes with fused P- and N-donor moieties.

Selected bond lengths for **8–Br** and **8–I** are given in Table 1. Comparison with reference compounds (see Table 2 and Table 3) shows that the Cu-halide and Cu-N bond lengths are similar to those in other Cu(I) complexes: The Cu-halide bonds are in the expected range, but will not be discussed further due to the rather larger variance (see Table 2). The Cu-N bonds are also in the expected range: While there are no known reference complexes with Cu(I) coordinated by phosphine and quinoline ligands, it can be seen that the Cu-N bond gets longer when moving from Cu(I) coordinated by pyridine to Cu(I) coordinated by pyridine and phosphine. The Cu-N bond for Cu(I) coordinated by quinoline (2.13 Å) is in a similar range as for Cu(I) coordinated by pyridine and phosphine.

However, the Cu-P and the P-O bond lengths deviate from the expected values: It is evident that the Cu-P bonds for both complexes are roughly 0.04–0.14 Å shorter than expected. While in most known cases, this bond is in the order of 2.19–2.29 Å, they are only 2.15/2.18 Å for **8–Br/8–I**, respectively. More anomalies can be found in the ligand backbone of the POQ unit. So far, only two other publications with POQ are available, studying the coordination chemistry with Ni(II)[77] as well as Fe(II) and Ru(II).[76] Nevertheless, the P-O bond seems to be slightly longer than expected (approx. 0.02–0.04 Å), while the C_{quin}-O bond is again close to the expected value.

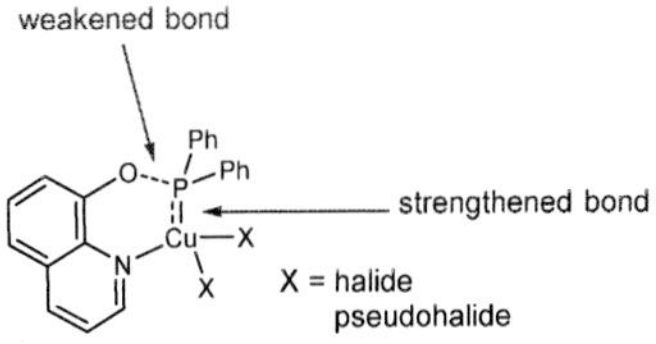

Figure 14: Under the hypothesis of differences in bond lengths hinting towards differences in binding strengths, the P-O bond is weakened upon coordination of POQ to Cu(I), whereas the Cu-P bond is stronger than in other complexes.

If the difference in bond lengths is considered to be a hint for strengthened (shortening) or weakening (elongation) of the respective bond, coordination of POQ to copper halides seems to weaken the bond between the PPh$_2$ unit and the quinolate unit, while the bond between PPh$_2$ and Cu(I) is stronger than usual. It is unlikely that the bond length anomalies are caused by steric hindrance and the presence of a chelating ligand. Usually, such effects are more likely to cause distorted bond angles, as it was found for other dinuclear copper(I) complexes.[29] The presence of elongated or shortened bonds has two important consequences: i) Upon photoexcitation, such bonds could be cleaved or open up non-radiative relaxation pathways and ii) the elongated bond could be more likely to be cleaved.

Various crystallization experiments with a Cu(I) precursor (CuCl, CuBr, CuI, and CuSCN) and POQ by diffusion of diethyl ether into acetonitrile solutions were performed. For complex **10**, triphenylphosphine was added. Single crystal samples of complexes **9–11** (Figure 15) were obtained.

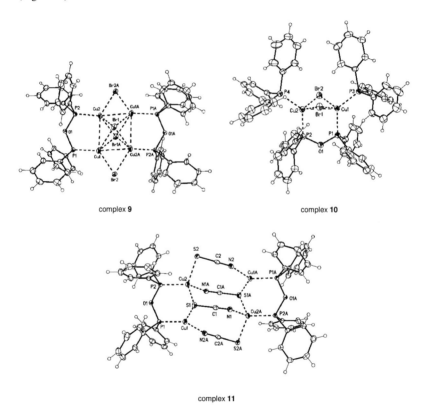

complex **9**   complex **10**

complex **11**

Figure 15: Molecular structures of complexes **9**, **10**, and **11**. Displacement parameter drawn at the 50% probability level.

All three isolated compounds did not contain the initial ligand POQ. Instead, a bridging bisphosphine ligand (PPh$_2$)O(PPh$_2$) (ligand **2**) was found.

Tetraphenyldiphosphoxane, (POP) is the tautomeric form of tetraphenyldiphosphine monooxide (PPO), which can be synthesized by hydrolysis of chlorodiphenylphosphine. Without the presence of coordinating metal ions, PPO is the dominating tautomer, while POP can be stabilized in the presence of transition metal ions. Early results for POP complexes were published in the 1970s with Cr, Mo, and W as metal centers.[79, 80, 82] Depending on the metal fragments, POP may coordinate in a bridging or chelating way, the latter yielding four-membered rings. Naktode and coworkers recently reported the first Cu(I) complexes with POP ligands,[81] which were obtained by the hydrolysis of N,N-bis-(diphenylphosphino)-aniline with copper halides in CH_2Cl_2/water mixtures. Interestingly, this procedure yielded not only POP complexes, but also multinuclear, heteroleptic complexes with POP as well as its isomer PPO, when additional triphenylphosphine was present.

The degradation only occurred when protic solvents such as acetonitrile or alcohols were present. From analysis of aged solutions of complex **8–I** in dichloromethane, non-dried acetonitrile as well as acetonitrile/water (v/v, 1/1) with gas chromatography, a transfer of oxygen from the quinolate moiety could be ruled out:

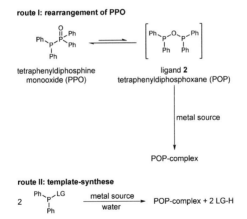

Scheme 2: Chemistry of the tetraphenyldiphosphoxane (POP) ligand.[79]

The formation of PPO and 8-hydroxyquinoline was observed, while neither quinoline, nor diquinolinyl ether or other coupling products were found. Trace amounts of water, which may be present in acetonitrile, destabilize the complexes by protonation of the quinoline moieties and serve as a source for the oxygen in the POP ligand. Furthermore, the oxidation of the coordinated ligand by O_2 does not seem to be of relevance for POQ Cu(I) system, given that complexes **8–Cl**, **8–Br**, **8–I** and **8–SCN** as powder samples were isolated at ambient conditions.

No formal mechanistic studies regarding the degradation pathway from a Cu(I) POQ to a Cu(I) POP species was performed. However, based on the herein presented results and similar reactions from the literature, a mechanism for the reaction can be proposed. While there is so far no concise analysis of a broad range of POP complexes in one single publication, the synthetic pathways proposed so far can be classified by either one of two groups (Scheme 2). Route 1 describes the thermal isomerization of PPO in the presence of a metal precursors, as reported by Wong and coworkers.[82] More common is route 2, where a phosphine precursor with a leaving group is employed. In this case, LG⁻ is the deprotonated form of 8-hydroxyquinoline, while Coetzee et al. used carboxylates as a leaving group and Rh as a metal center.[83] Zeiher and coworkers used sulfonates in combination with Cr, Mo and W,[84] while Renz and coworkers used 3-hydroxypyridine and Ag.[85] Older works describe the use of chlorodiphenylphosphine with chloride as a leaving group as precursor.[86, 87]

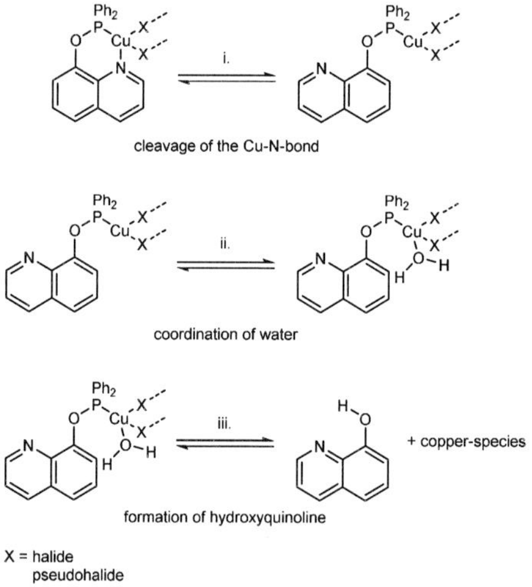

cleavage of the Cu-N-bond

coordination of water

formation of hydroxyquinoline

X = halide
pseudohalide

Scheme 3: Proposed mechanism for the degradation of Cu(I) POQ complexes.

According to route II, the mechanism depicted in Scheme 3 can be proposed for the herein reported Cu(I) POQ complexes: It can be speculated that the first, rate determining step (i) is the cleavage of the Cu-N bond. From the molecular structures of **8-Br** and **8-I**, it is evident that the bond between Cu and the N in quinoline is not weakened. After a 16 valence electron species is formed, water may coordinate (ii) and transfer a proton to the quinolate, which allows for the cleavage of the P-O bond. This step (iii) yields a Cu(I)(OH)PPh$_2$ species, which can react with an additional equivalent of POQ to form POP.

In conclusion, a new approach to synthesize red-emitting Cu(I) complexes with chelating ligands by covalent connection of two monodentate ligands was established. However, the materials seem to be prone to degradation in the presence of water. Based on the analysis of differences in bond lengths, obtained by single crystal x-ray diffraction, and chemical analysis of the degraded solution of the Cu(I) POQ complexes, a hypothesis for the degradation mechanism was developed.

For the application as emitters, the degradation reaction found in the POQ complexes needs to be suppressed in modified complex structures in a future development step. So far, only limited knowledge and hypotheses exist regarding the degradation of Cu(I) emitters in OLEDs. It was demonstrated that ligands that contain potential leaving groups may lead to degradation of the emissive complex. Furthermore, strained or compressed bonds in the emissive complex may open up non-radiative decay pathways and again lead to disposition for degradation.

4.1.2 Red-emitting Cu(I) complexes with alkyne-substituted ligands

Cu(I) complexes bearing alkyne-substituted ligands can undergo intramolecular-catalyzed, Cu(I)-catalyzed azide alkyne cycloadditions (CuAAc). This reaction concept was used to link the green-emitting NHetPHOS complex **7** to an azide-substituted polymer as well as benzyl azide.[88] The structure of NHetPHOS complex **7** is in accordance with other NHetPHOS complexes. The complex consists of a butterfly-shaped Cu$_2$I$_2$ core and three butynylPyrPHOS ligands, one which is bridging the two Cu(I) centers and two which are terminal. Each Cu(I) atom features a pseudo-tetrahedral coordination environment (see Figure 16).[88]

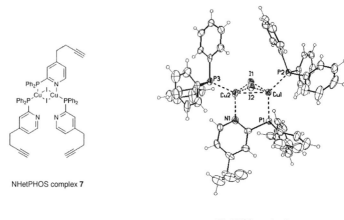

NHetPHOS complex **7**

NHetPHOS complex **8**

Figure 16: Molecular structures of NHetPHOS complex **7** and NHetPHOS complex **8**,[88, 89] the latter was derived by singly crystal x-ray diffraction.

In analogy, the heteroleptic NHetPHOS complex **8** containing one bridging butynylPyrPHOS ligand and two terminal triphenylphosphine ligands was reported.[89]

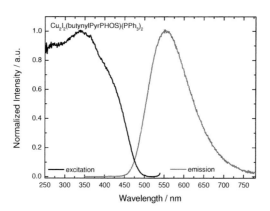

Figure 17: Excitation ($\lambda_{Em.}$ = 550 nm) and emission ($\lambda_{Ex.}$ = 340 nm) spectra of powder samples of NHetPHOS complex **8**.

Single crystals suitable for single crystal x-ray diffraction (XRD) of NHetPHOS complex **8** were obtained in context of this work (see Figure 16) and optical excitation and emission spectra were measured. The peak emission intensity for powder samples of NHetPHOS complex **8** was determined to be at 554 nm (see Figure 17), which is very similar to the peak emission intensity for powder samples of NHetPHOS complex **7** at 548 nm.[88]

It can be advantageous to immobilize the Cu(I) emitter on a host polymer in a solution-processed OLED as blending of the EL with neighboring layers can be prevented that way without the requirement for orthogonal solvents.[88] To extend the family of NHetPHOS complexes with alkyne-substituted ligands, known ligands, i.e., butynylPyrPHOS and butynylPHOS were chosen and two heteroleptic Cu(I) complexes were synthesized. NHetPHOS complex **9** consists of a Cu_2Br_2 core and contains a bridging butynylPyrPHOS and two terminal tris(o-methoxyphenyl)phosphine ligands (see Figure 18). ButynylPyrPHOS was synthesized according to an established method.[88] It has been demonstrated that the optical emission is shift to longer wavelengths for NHetPHOS complexes containing bromine or chlorine instead of iodine as halide. However, the peak of the optical emission intensity of Cu_2Br_2(2-(diphenylphosphino)pyridine)$_3$ is only shifted by 14 nm to longer wavelengths in comparison to Cu_2I_2(2-(diphenylphosphino)-pyridine)$_3$.[48] Due to the similarity of the molecular structure of Cu_2Br_2(2-(diphenylphosphino)pyridine)$_3$, it can be assumed that NHetPHOS complex **9** also features a yellow-green emission. To obtain a potentially red-emitting NHetPHOS complex with alkyne-substituted ligands, NHetPHOS complex **10** consisting of a Cu_2I_2, a bridging IsoquinPHOS ligand, and two terminal butynylPHOS ligands was synthesized (see Figure 18).

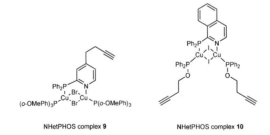

NHetPHOS complex **9**   NHetPHOS complex **10**

Figure 18: Molecular structures of NHetPHOS complex **9** and NHetPHOS complex **10**.

IsoquinPHOS[90] and butynylPHOS[89] were synthesized according to established procedures and provided by Dr. Daniel Volz. Cu_2I_2(1-(diphenylphosphino)isoquinoline)$_3$

exhibits a peak emission intensity at 657 nm.[48] It can be assumed that NHetPHOS complex **10** also features an orange-red emission.

In addition to the dinuclear NHetPHOS complexes, mononuclear Cu(I) complexes with alkyne-substituted ligands were developed. Phenanthroline and its derivatives were chosen as ligands, as numerous cationic Cu(I) phenanthroline and 2,9-dimethyl-phenanthroline (dmp) complexes exhibiting a red emission were reported. In these complexes the Cu(I) atom is either coordinated by two chelating phenanthroline-type ligands or one chelating phenanthroline-type ligand and two monodentate ligands, e.g., phosphines.[91-93] The charge is balanced by a non-coordinating anion, e.g., BF_4^- or PF_6^-. These complexes are referred to as cationic. For most of these complexes, photophysical data for solid sample were not reported. Therefore, a mononuclear, cationic Cu(I) complex with one phenanthroline and two triethyl phosphite ligands (complex **12**, see Figure 19) was synthesized in order to determine its photoluminescence quantum yield (PLQY) in solid state. The luminescence of complex **12** was weak with a low photoluminescence quantum yield of PLQY = 0.010 in solid state.

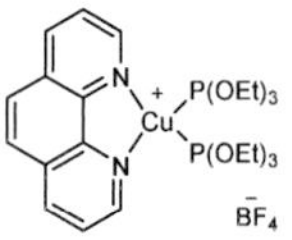

complex **12**

Figure 19: Molecular structure of complex **12**.

Besides cationic, mononuclear Cu(I) complexes, there are also a few examples for neutral, mononuclear Cu(I) complexes with phenanthroline or derivatives of phenanthroline as ligands, in which the positive charge is balanced by a coordinating halide.[35, 94-97] Two neutral, mononuclear Cu(I) complexes with alkyne substituents were synthesized. Complex **13** contains iodine as counter ion, one dmp ligand, and one butynylPHOS ligand (Figure 20). An alkyne-substituted phenanthroline-derivative butynyl-Me-phen was developed to allow for variation of the structural motif and deployment of other phosphorus ligands. Therefore, 2,9-dimethyl-phenanthroline was deprotonated with lithium diisopropylamide (LDA) and reacted with propargyl bromide similar to an established procedure.[98] Upon conversion of butynyl-Me-phen and triphenylphosphine with CuI, complex **14** (see Figure 20) was obtained and single crystals suitable for single crystal x-ray diffraction were grown.

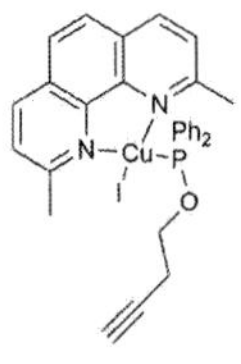

complex **13**

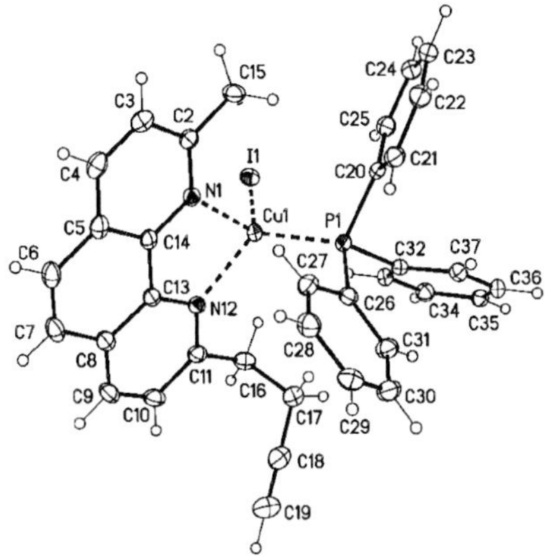

complex **14**

Figure 20: Molecular structures of complex **13** and complex **14**, the latter was derived by single crystal x-ray diffraction.

Optical excitation and emission spectra of complex **14** in solid state as well as in solution (dichloromethane) are depicted in Figure 21. The features in the excitation and emission spectra are broad and unstructured as for other Cu(I) complexes.

The low energy onset of the excitation feature is shifted by approximately 40 nm and the peak maximum of the emission by approximately 105 nm to lower energies for the solution spectra in comparison to the solid state spectra. Whereas the excitation feature is less broad in solution the emission feature is broader in solution. The reason for this behavior, which also occurs in other Cu(I) complexes, in not yet understood. Dissociation, ligand exchange reactions, influences of the molecular environment or different excited state dynamics might play a role. In solid state complex **14** exhibits a photoluminescence quantum yield of PLQY = 0.035.

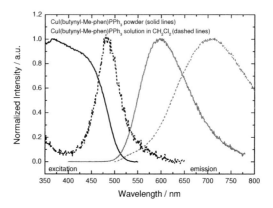

Figure 21: Excitation ($\lambda_{Em.}$ = 615, 640, 630 nm) and emission ($\lambda_{Ex.}$ = 350 nm) spectra of powder and solution (CH$_2$Cl$_2$) samples of CuI(butynyl-Me-phen)PPh$_3$.

To demonstrate that complex **14** can undergo an intramolecular-catalyzed CuAAc, it was reacted for several hours with three equivalents benzyl azide in DMSO. FAB-MS of the crude product was measured, which showed conversion of the alkyne and azide to a triazole (see Figure 22). The m/z peaks at 479/669 can be assigned to triazole-Me-phen, from which benzyl groups were cleaved, associated with CuI/Cu$_2$I$_2$.

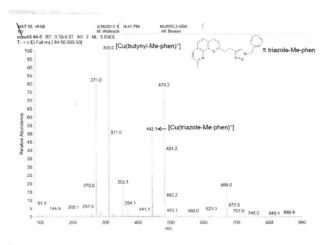

Figure 22: FAB-MS spectrum of the crude reation product of complex **14** with benzyl azide in DMSO after several hours.

In the context of this work, two potentially red-emitting Cu(I) complexes with alkyne-substituted ligands were synthesized. NHetPHOS complex **10** consist of a Cu_2I_2 core, a bridging IsoquinPHOS ligand, and two terminal butynylPHOS ligands. The neutral, mononuclear complex **13** contains iodine as counter ion, one dmp ligand, and one butynylPHOS ligand. To deploy these complexes as immobilized emitters in an OLED the photophysical properties of these complexes should be characterized and it should be tested whether these complexes undergo an intra-molecular catalyzed CuAAc. Furthermore, an alkyne-substituted phenanthroline derivative was developed and the neutral, mononuclear complex **13** containing iodine as counter ion, one triphenylphosphine ligand, and one butynyl-Me-phen ligand was synthesized. This complex exhibits an emission maximum at $\lambda_{max} = 597$ nm and a moderate photoluminescence quantum yield of PLQY = 0.035 and can undergo an intra-molecular catalyzed CuAAc. Therefore, this complex should be tested as an immobilized emitter in an OLED.

4.2   Molecular structure

4.2.1 Critical review of different methods for the assessment of the molecular structure of Cu(I) emitters in OLEDs

4.2.1.1 Single crystal x-ray diffraction, photophysical measurements, IR and NMR spectroscopy

For the application in OLEDs, processing of Cu(I) NHetPHOS complexes from solution, e.g., with coating or printing techniques is advantageous. For this purpose, the emitter is processed from bulk powders or crystals to amorphous thin films with a matrix material containing the emitter. Whereas the deployed crystalline material can be characterized by single crystal x-ray diffraction, confirming the preservation of the molecular structure in the amorphous thin film upon solution-processing is challenging. The comparison of photophysical emission spectra of the deployed bulk material and the operating device can only give a first hint whether a change in the molecular structure has occurred. As the emission is stimulated by UV or visible light for the former and by an electronic field for the latter, different electronic processes might lead to the emission of visible light. Furthermore, as the morphology, packing, and dipole moment of the environment in the bulk and the thin film is different due to possibly residual solvent molecules and the matrix material,[99] changes in the shape and position of the visible emission band can be expected even if the molecular structure is preserved upon processing. In case of the formation of different species upon processing, e.g., tetranuclear Cu(I) NHetPHOS complex with a metal-to-ligand ratio of 1:1 and residual, excess ligand instead of dinuclear Cu(I) NHetPHOS complexes with a metal-to-ligand ratio of 2:3, there would not be a detectable difference found in elemental analysis and the change in the vibrational bands in the IR spectra might be not significant, especially since the shape of many IR spectra is dominated by ligand modes, while copper-halide modes are usually not detectable in the standard range of this method (4000 to 400 cm^{-1}). It previously was reported that the characterization of Cu(I) NHetPHOS complexes by NMR can be problematic,[100, 101] mainly due to a plethora of NMR-active nuclei, e.g., ^{31}P $\left(\text{nuclear spin } S = \frac{1}{2}\right)$, ^{63}Cu $\left(S = \frac{3}{2}\right)$, ^{65}Cu $\left(S = \frac{3}{2}\right)$, and ^{127}I $\left(S = \frac{5}{2}\right)$. As the NMR-active copper and iodine nuclei exhibit a nuclear spin greater $S = \frac{1}{2}$ and a quadrupolar momentum, the nuclear energy levels split upon the application of a magnetic field according to $2S + 1$. The NMR signals of quadrupolar nuclei are usually wider than those with $S = \frac{1}{2}$ due to rapid quadrupolar relaxation. This may lead to broad, unstructured signals with complex multiplet coupling, even when studying solid samples with MAS-NMR techniques.[102]

X-ray absorption spectroscopy offers an alternative method to characterize Cu(I) NHetPHOS complexes: XAS with hard energetic x-ray radiation is a well-established method for the characterization of the coordination environment of metal atoms in non-crystalline materials, e.g., to investigate the composition of metal-based paint in historic paintings,[103] the coordination environment of metal centers in metal-containing proteins,[104] and metal-rich mining waste in environmental samples.[105] This method was previously used to characterize the molecular structure of Cu(I) complexes in amorphous thin films formed by co-deposition of a ligand and CuI[106, 107] and is also suitable for the investigation of amorphous thin film samples of the luminescent Cu(I) NHetPHOS complexes. In this work, Cu K x-ray absorption spectroscopy was used to investigate the molecular structure of Cu(I) NHetPHOS complexes in crystalline, amorphous powder, thin film, and solution samples. For a comprehensive investigation of the materials solid-state, CP/MAS ^{31}P NMR on crystalline and amorphous solids and thin films was used in cooperation with Dr. Stephan L. Grage and Prof. Dr. Anne S. Ulrich as complementary technique. Using this approach, the local electronic structure around the phosphorus atoms and the local electronic and geometrical structure around the copper metal centers of amorphous and crystalline solids as well as thin film samples are probed with solid state ^{31}P-NMR and x-ray absorption spectroscopy (XAS) at the Cu K edge, respectively.

4.2.1.2 X-ray absorption spectroscopy

In x-ray absorption spectroscopy, the sample is irradiated by x-ray photons of a defined energy and the incident x-ray energy is scanned across the absorption edge. Some of the x-ray photons are absorbed by the atoms in the sample and core electrons are excited or ejected in dependency of the energy of the incident x-ray photons. The system is excited from its initial state $|\Psi_i\rangle$ into the final state $\langle\Psi_f|$. The x-ray absorption process is referred to as resonant if the electron is excited into an unoccupied state and as non-resonant if the electron is ejected. For resonant x-ray absorption, the interaction of the core hole with the excited electron (core exciton) has to be taken into account. The absorption intensity can be determined by measuring the intensity of the incident beam and the transmitted beam, by measuring the x-ray fluorescence emitted by valence electrons filling the core hole, or by measuring the Auger electrons, which are ejected as the core hole is filled.

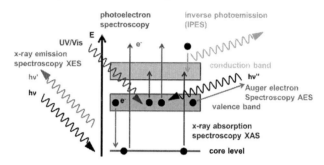

Figure 23: Schematic illustration of x-ray spectroscopic techniques as well as UV/Vis, ultraviolet, and inverse photoemission spectroscopy.

The probability for the absorption of a photon can be described according to Fermi`s Golden Rule as follows:[108, 109]

$$\omega_{i\to f} \propto |\langle \Psi_f | \hat{H}_s | \Psi_i \rangle|^2 \, \delta(E_f(N) - E_i(N) - h\nu) \tag{4.1}$$

$|\Psi_i\rangle$ and $\langle \Psi_f|$ represent the total wave function of the initial and final state, respectively. $\hat{H}_s$ represents the perturbation of the system by the electromagnetic field of the photon.[109]

$$\hat{H}_s = -\frac{e}{2m_e c}(\hat{p}\hat{A} + \hat{A}\hat{p}), \text{ with } \hat{A}(\vec{r}, t) = \vec{e}\hat{A}_0 e^{i(\vec{k}\vec{r} - \omega t)}, \hat{p} = -i\hbar\vec{\nabla} \tag{4.2}$$

$\vec{e}$ is the polarization vector of the electromagnetic wave. $\hat{A}$ can be simplified by applying the dipole approximation, in which only the first term of the series expansion of the exponential function is taken into account. The dipole approximation only applies if the wavelength of the photon $\vec{k}\vec{r}$ is much larger than the extent of the atomic wave function, which is typically assumed (but not strictly valid) for photon energies up to 1 keV. By introducing the Coulomb gauge $(\hat{p}\hat{A} = 0)$, the dipole operator can be written as $\hat{H}_s = -\frac{e}{2m_e c}\vec{e}\hat{p} = \vec{e} \cdot \vec{r}$.

In the solid state, the probability for the absorption of a photon of a certain energy can be derived from Fermi`s Golden Rule by substituting the δ-function with the local, partial density of states (DOS) of the unpopulated states ρ_{lp}.

$$A\left(h\nu_{absorption}\right) \propto \left|\langle \Psi_f | H_s | \Psi_i \rangle\right|^2 \rho_{lp}(E_i) \; with \; (E_i = E_f + h\nu_{absorption}) \qquad (4.3)$$

As the core hole generated upon excitation is strongly localized, the transition can only occur into local unpopulated states and therefore, the local, partial density of states ρ_{lp} can be deployed.

The transition matrix element ensures that there is a wave function overlap between the core electronic state and the final state. The transition has to fulfill the dipole selection rules for the initial and final state: $\Delta L = \pm 1$; $\Delta S = 0$; $\Delta J = \pm 1, 0$, except $J = 0 \leftrightarrow 0$; $\Delta m_J = \pm 1, 0$, except $m_J = 0 \leftrightarrow 0$ if $\Delta J = 0$. The symbol $\leftrightarrow$ is used to indicate that this transition is not allowed. $J = L + S$ is the total angular momentum. L is the azimuthal quantum number, S is the spin quantum number, and m_J is the magnetic quantum number. Additionally, the parity of the wave function has to change, which means that the symmetry of the wave function upon reflection at a mirror plane has to change.

4.2.2 Results: X-ray absorption spectroscopy at the Cu K edge

4.2.2.1 Molecular structure of three NHetPHOS complexes in single crystal, amorphous powder and thin film samples[†]

X-ray absorption spectroscopy at the Cu K edge was used to study whether the molecular structure of three Cu(I) NHetPHOS complexes is retained when preparing thin film from solution. This question is relevant when it comes to solution processing of the emissive layer of OLEDs. Significant differences in the optical emission spectra of amorphous solid,

---

[†] The data and analysis reported in this chapter was previously published in the context of this work. Figures, tables, and wording is reprinted and adapted where appropriate with permission from ([102]   D. Volz, M. Wallesch, S. L. Grage, J. Göttlicher, R. Steininger, D. Batchelor, T. Vitova, A. S. Ulrich, C. Heske, L. Weinhardt, T. Baumann, S. Bräse, *Inorg. Chem.* **2014**, *53*, 7838-7847.) Copyright (2014) American Chemical Society. The photophysical measurements and the respective analysis were performed by Dr. Daniel Volz and the MAS ^{31}P NMR experiments and their analysis were conducted by Dr. Stephan L. Grage and Prof. Dr. Anne S. Ulrich.

thin film, and solution samples of the investigated materials raise the question whether these Cu(I) complexes are stable upon processing of the EL from solution.[102] Therefore, Cu K XAS was measured on single crystal, amorphous bulk, and neat thin film samples of these complexes. X-ray absorption spectroscopy (XAS) at the Cu K edge provided insight into the local electronic and geometrical environment of the Cu(I) metal centers of the samples.[102] The Cu K XAS data was complemented by nuclear magnetic resonance (MAS ^{31}P NMR) to allow for an investigation whether the structure or even the stoichiometry of the complexes in the thin films samples is in agreement with structure obtained from x-ray diffraction on single crystals. The photophysical measurements and the respective analysis were performed by Dr. Daniel Volz and the MAS ^{31}P NMR experiments and their analysis were conducted by Dr. Stephan L. Grage and Prof. Dr. Anne S. Ulrich.

"All complexes were synthesized according to the procedures given in the references.[28, 29, 99] The crystalline samples were made by growing single crystals from dichloromethane/ether. Amorphous powders were made by precipitation from dichloromethane in methanol. Neat film samples were obtained by spin-coating of concentrated solutions of the compounds on quartz glass substrates (1×1 cm^2) in air."[102] "The structures of all complexes investigated herein were previously confirmed with single crystal x-ray diffraction.[29, 48, 99],[102] The molecular structures derived from this analysis are given in Figure 24, while selected distances and angles are displayed in Table 5. The elemental analysis for carbon, nitrogen, and hydrogen of crystalline, amorphous powder, and thin film samples for NHetPHOS complexes **1–3** confirmed that the composition of the samples is consistent with the empirical formula within the accuracy of this method.[102]

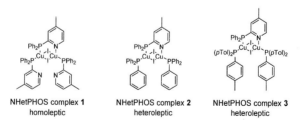

Figure 24: Molecular structures of NHetPHOS complexes **1–3**, which were studied by Cu K XAS, photophysical measurements, and MAS ^{31}P NMR.†

Comparing the emission spectra measured on different samples (powders, thin films, and solutions) several differences can be found (Figure 25 and Table 4): The film spectra feature a red shift relative to the powder spectra ($\Delta\lambda$ = 31, 31, and 11 nm for NHetPHOS complexes **1–3**), while the emission band is slightly broadened.[102] In solution, this trend continues with even broader bands and higher red-shifts in comparison to the powder samples ($\Delta\lambda$ = 66, 62, and 42 nm for NHetPHOS complexes **1–3**).[102]

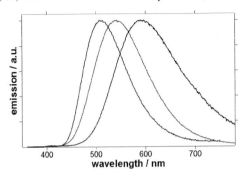

Figure 25: Emission spectra of NHetPHOS complex **2** – from left to right – as a powder, a neat thin film, and in solution. Excitation at 350 nm.[102]†

Furthermore, the photoluminescence quantum yield (PLQY) drastically differs for amorphous powder, thin film, and solution samples. Whereas the PLQY for amorphous powder samples is in the range of 74%–99%, the PLQY for solution samples is reduced to 2% to 24% (Table 4).[102] Comparing amorphous powder, thin film, and solution samples there is no clear trend for the changes in the optical emission decay time τ (Table 4).[102]

"Earlier attempts to explain such behavior refer to a rigidochromic effect, first used to explain differences in emission spectra of Cu(I) complexes between solid and dissolved samples.[110] The observation that differences are also found when comparing thin films with bulk material raises the question whether this particular red shift is really a result of morphological effects like the rigidity or electronic structure of the environment of the excited chromophores, or whether other effects like the actual formation of other species must be considered."[102]

Table 4: Photophysical properties of NHetPHOS complexes **1–3** for powder, solution, and film samples.[28, 48, 99, 102]†

		NHetPHOS complex **1**	NHetPHOS complex **2**	NHetPHOS complex **3**
powder	λ_{max} / [nm]	515	510	542
	PLQY Φ	0.89	0.99	0.74
	$<\tau>$ / [µs]	2.3	1.9	3.7
solution	λ_{max} / [nm]	581	572	589
	PLQY Φ	0.02	0.24	0.14
	$<\tau>$ / [µs]	0.2	2.4	1.7
film	λ_{max} / [nm]	546	541	553
	PLQY Φ	0.92	0.94	0.92
	$<\tau>$ / [µs]	3.7	2.3	2.2

λ_{max}: emission (peak maximum, excitation at 350 ± 1 nm); PLQY: excitation at 350 nm, Φ error bar: ± 0.02; $<\tau>$ error bar: ± 0.4 µs. Emission decay times were fitted using the intensity average lifetime as proposed by O'Connor and coworkers.[111] Solutions were measured in degassed toluene at 10 mg mL^{-1} at room temperature. Films were made by spin-coating from toluene onto quartz substrates, typical thicknesses was 20 nm.[102]

Cu K XAS data were analyzed to investigate whether the structure and the stoichiometry of the complex in the thin films samples is in agreement with the structure obtained from x-ray diffraction on single crystals.

Table 5: Selected bond lengths and angles for NHetPHOS complexes **1–3** derived from single crystal-ray diffraction analysis.[102]†

	NHetPHOS complex **1**	NHetPHOS complex **2**	NHetPHOS complex **3**
lengths			
Cu–Cu	2.753(1)	2.8164(5)	2.7135(7)
Cu$_P$–I	2.682(1) 2.688(1)	2.6864(5) 2.6878(4)	2.6871(6) 2.6535(7)
Cu$_P$-P$_{P,N}$	2.247(2)	2.2353(8)	2.250 (1)
Cu$_N$-N	2.078(1)	2.090(2)	2.105(3)
Cu$_N$-P$_P$	2.253(1)	2.2533(7)	2.252 (1)
Cu$_P$-P$_P$	2.247(1)	2.2493(7)	2.250 (1)
angles			
Cu–I–Cu	62.34(1) 61.85(1)	63.953(1) 63.357(1)	60.837(2) 61.213(2)
P–Cu–P	118.47(1)	123.90(3)	121.20(4)

Bond lengths are given in Å, angles are in degree (°).

In Figure 26, XANES spectra for crystalline samples of NHetPHOS complexes **1**, **2**, **3**, and powder sample of NHetPHOS complex **2** are depicted. From the XANES region, information on the oxidation state as well as on the coordination number and symmetry of the copper ion can be extracted. Cu(I) complexes typically show low energy peaks in the region between 8983.0 eV and 8986.0 eV and one peak at 8990.0 eV.[104] "The former peaks are assigned as $1s \rightarrow 4p_x$ and $1s \rightarrow 4p_{y,z}$ transitions.[112] For the samples presented in this study these peaks were observed at 8984.0 eV and 8990.5 eV."[102]

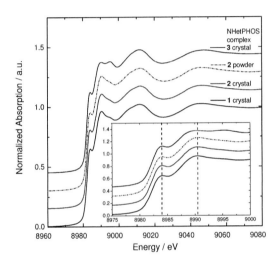

Figure 26: Cu K XANES spectra for crystalline samples of NHetPHOS complexes **1**, **2**, and **3** and powder sample of NHetPHOS complex **2**. "Spectra are calibrated in energy (the inflection point of the edge of a Cu metal reference foil is set to 8980.3 eV[112] and normalized to edge jump). Vertical lines at 8984.0 and 8990.5 eV mark prominent features in the spectra. The inset shows the edge region in more detail."[102]†

While the spectra of NHetPHOS complexes **1** and **2** exhibit a shoulder in the range between 8994.0 and 8997.0 eV in crystalline form, NHetPHOS complex **3** exhibits another well-defined peak at 8995.3 eV.[102] The XANES spectra of NHetPHOS complex **2** of the crystalline and amorphous powder samples are very similar.[102] In contrast to Cu(I) complexes, Cu(II) complexes show intense peaks in the region between 8986.0 eV and 8988.0 eV, which are assigned to $1s \rightarrow 4p$ transitions, and exhibit a pre-edge peak at

8979.0 eV that corresponds to 1s→3d transitions.[112] Cu(I) NHetPHOS complexes **1–3** do not shows these features, which confirms that the d orbitals are fully occupied and that the oxidation state of copper is +1.[113] "The energy position of the 1s→4p transitions in Cu(I) complexes is correlated to the coordination number of Cu(I) and the symmetry of the ligands around the central atom.[104, 114],[102] The degenerated $p_{x,y,z}$ orbitals of a Cu(I) ion in tetrahedral coordination split according to the ligand field theory upon distortion of the local geometry and changes to three- and twofold coordination.[104, 112] The energies of the 1s→4p transitions (8984.0 and 8990.5 eV) are in agreement with the energies reported in literature for other Cu(I) iodide complexes and with the distorted tetrahedral coordination of copper in these complexes.[106, 115]

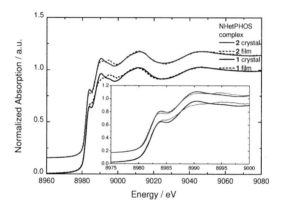

Figure 27: Cu K XANES spectra of NHetPHOS complexes **1** and **2** in crystalline form and as thin film. Spectra are calibrated in energy and normalized to edge jump. The inset shows the edge region in more detail.[102]†

In Figure 27, a comparison of the crystalline and film samples of NHetPHOS complexes **1** and **2** is given. Single crystals obtained from the same batch were used to resolve the structure of complex **1** by single crystal x-ray diffraction, so they serve as an additional reference with a known structure.[102] The primary features in the spectra of the crystalline form are all reproduced in the film spectra, which indicates that there is no drastic change in coordination geometry.[102] However, for both complexes there is a "smearing out" of the first and second resonances and an additional feature around 8995.3 eV, as was also found in the spectrum of NHetPHOS complex **3** in crystalline form. "A possible explanation may be a small change in the local geometry of the complex upon processing of thin films from solution. Further information on the local environment of the Cu(I) centers can be drawn from the analysis of the EXAFS region (Figure 28 and Figure 29). Whereas from the electronic structure reflected in the XANES spectra, information on the molecular geometric properties of the absorption site can be extracted, EXAFS can be used to derive nearest neighbor distances."[102] Figure 28 presents EXAFS data in k space ($k^3\chi(k)$) for crystalline samples of NHetPHOS complexes **1–3** and the powder sample of NHetPHOS complex **2**, as well as the amplitude of the Fourier transformed $k^3\chi(k)$ ($FT[k^3\chi(k)]$) signal in R space. No significant differences are detectable between the EXAFS data of crystalline samples of NHetPHOS complexes **1–3** and the amorphous powder sample of NHetPHOS complex **2**, respectively.[102] Furthermore, the EXAFS data of crystalline, amorphous, and film samples of NHetPHOS complexes **1** and **2** are in good agreement with each other.[102] The comparison of EXAFS data $k^3\chi(k)$ and $FT[k^3\chi(k)]$ of a crystalline and a thin film sample of complexes **1** and **2** are shown in Figure 29.

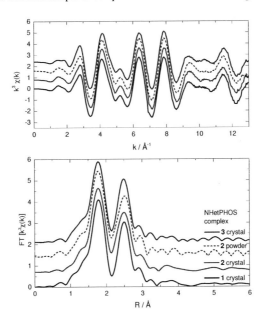

Figure 28: EXAFS data $k^3\chi(k)$ (top) and $FT[k^3\chi(k)]$ (bottom) of crystalline samples of NHetPHOS complexes **1–3** and a powder sample of complex **2**.[102]

"Interatomic distances and coordination numbers (Cu-N, Cu-P, Cu-I and Cu-Cu) derived from the single crystal x-ray diffraction analysis of the corresponding crystalline samples were used to set up the starting structures. One Cu-N, Cu-P, Cu-I, and Cu-Cu single scattering path was included to obtain the best fits."[102]

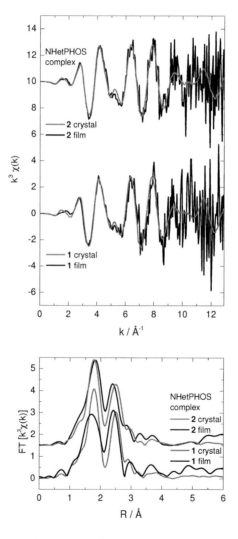

Figure 29: Comparison of EXAFS data $k^3\chi(k)$ (top) and $FT[k^3\chi(k)]$ (bottom) of a crystalline and a thin film sample of NHetPHOS complexes **1** and **2**.[102]†

As an example the best fits for a crystalline and a film sample of NHetPHOS complex **1** are shown in Figure 30. The obtained structural data is collected in Table 5. "For all samples, the coordination numbers are in reasonable agreement with the mean coordination numbers (N) of the two copper centers derived from the crystal structure (N(N) = 0.5, N(P) = 1.5, N(I) = 2.0, and N(Cu) = 1.0), given the high correlation between coordination numbers and Debye-Waller factors in particular in view of the rather high Debye-Waller factors found for the Cu-Cu path. The interatomic distances Cu-N and Cu-P are in good agreement, whereas the Cu-I distances are about 0.1 Å shorter and the Cu-Cu distances are about 0.2 Å longer obtained from EXAFS compared to single crystal diffraction results. Such differences are possible due to the sensitivity of the EXAFS spectrum to the mean short range atomic order around the absorbing atom, whereas long range order is measured in single crystal diffraction."[102]

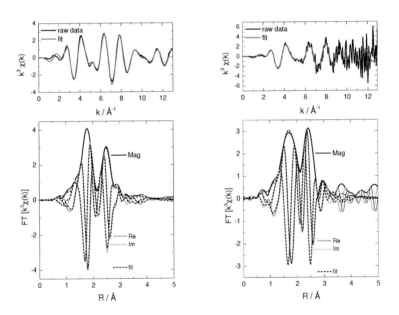

Figure 30: EXAFS data $k^3\chi(k)$ (top) and FT[$k^3\chi(k)$]: magnitude (Mag), real (Re) and imaginary (Im) parts (bottom) and their best fits for a crystalline (FT range k = 2.3 - 12.2 Å^{-1}) sample of NHetPHOS complex **1** (left) and for a and a thin-film sample (FT range k = 2.3 to 11.2 Å^{-1}) of NHetPHOS complex **1** (right).[102]†

"The peak between 1.0 and 2.1 Å (not phase-shift corrected) is modeled with Cu-N, Cu-P and Cu-I scattering paths. The peak between 2.1 and 3.4 Å (not phase-shift corrected) describes the scattering of the photoelectron from I and Cu atoms (Cu-I and Cu-Cu). The Cu-I single scattering path has a complex shape and contributes in both peaks (see Figure 31)."[102]

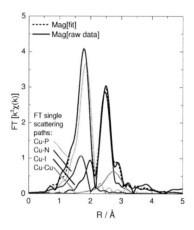

Figure 31: EXAFS of crystalline sample of NHetPHOS complex **1**: magnitude of the FT[$k^3\chi(k)$], best fit and single scattering paths used in the fit.[102]†

"The amplitude reduction factor $S_0^2 = 0.9$, which accounts for multi-electronic excitations, is obtained from a fit of an EXAFS spectrum of a metallic Cu foil measured in the same experimental conditions. S_0^2 is fixed to 0.9 for all fits."[102] The values for the energy threshold offset ΔE_0 extracted from the fits are 0.6 ± 0.3 eV for the crystalline sample of NHetPHOS complex **1** and −0.6 ± 0.4 eV for the thin-film sample of NHetPHOS complex **1**.[102]

"The Debye-Waller factors σ^2 derived from the model fit lie between 0.005 and 0.026 Å², with increasing values corresponding to longer interatomic distances. The relatively high values for the Cu-Cu path can be attributed not only to thermal disorder at room temperature,[116, 117] but also suggest a static disorder contribution.[118] Comparing the fitting results for crystalline and film samples shows that there are no significant differences between the spectra as the results for ΔE_0, N, R, and σ^2 are very similar within the error bars."[102] Fitting results for crystalline and film samples of NHetPHOS complex

1, for crystalline, amorphous powder and film samples of NHetPHOS complex **2**, and crystalline sample of NHetPHOS complex **3** are presented in Table 6.

Table 6: EXAFS results for crystalline and film samples of NHetPHOS complex **1**, for crystalline, amorphous powder, and film samples of NHetPHOS complex **2***, and crystalline sample of NHetPHOS complex **3**. From left to right: single scattering paths used in the fit (path), interatomic distances (R), coordination numbers (N), deviation from the distances obtained from the XRD analyses, mean squared atomic displacement/Debye-Waller factors (σ^2). The N(Cu) is restrained during the fit N(I)+N(Cu)=3.[102]†

The value for ΔE_0 extracted from this fit is 0.6 ± 0.3 eV.				
NHetPHOS complex **1** (crystalline), R-factor = 0.002				
path	R / Å	N	ΔR / Å	σ^2 / Å^2
Cu-N	2.10	0.6 ± 0.1	0.01 ± 0.01	0.006 ± 0.000
Cu-P	2.25	1.1 ± 0.1	0.01 ± 0.01	0.006 ± 0.000
Cu-I	2.64	1.4 ±0.1	−0.05± 0.01	0.011 ± 0.001
Cu-Cu	3.05	1.6	0.23 ± 0.02	0.021 ± 0.003

The value for ΔE_0 extracted from this fit is −0.6 ± 0.4eV.				
NHetPHOS complex **1** (film), R-factor = 0.001				
path	R / Å	N	ΔR / Å	σ^2 / Å^2
Cu-N	2.10	0.9 ± 0.2	−0.01 ± 0.01	0.008 ± 0.001
Cu-P	2.25	1.1 ± 0.2	−0.01 ± 0.01	0.008 ± 0.001
Cu-I	2.64	1.2 ± 0.1	−0.09 ± 0.01	0.008 ± 0.001
Cu-Cu	3.05	1.8	0.27 ± 0.04	0.026 ± 0.006

The value for ΔE_0 extracted from this fit is 0.4 ± 0.3 eV.				
NHetPHOS complex **2** (crystalline), R-factor = 0.003				
path	R / Å	N	ΔR / Å	σ^2 / Å^2
Cu-N	2.10	0.4 ± 0.1	0.01 ± 0.01	0.005 ± 0.000
Cu-P	2.25	1.1 ± 0.1	0.01 ± 0.01	0.005 ± 0.000
Cu-I	2.62	1.4 ±0.1	−0.06 ± 0.01	0.011 ± 0.001
Cu-Cu	3.03	1.6	0.22 ± 0.02	0.019 ± 0.003

Table 6: EXAFS results for crystalline and film samples of NHetPHOS complex **1**, for crystalline, amorphous powder, and film samples of NHetPHOS complex **2**[*], and crystalline sample of NHetPHOS complex **3**. From left to right: single scattering paths used in the fit (path), interatomic distances (R), coordination numbers (N), deviation from the distances obtained from the XRD analyses, mean squared atomic displacement/Debye-Waller factors (σ^2). The $N(\text{Cu})$ is restrained during the fit $N(\text{I})+N(\text{Cu})=3$.[102]†

The value for ΔE_0 extracted from this fit is 0.6 ± 0.3 eV.				
NHetPHOS complex **2** (powder), R-factor = 0.002				
path	R / Å	N	ΔR / Å	σ^2 / Å^2
Cu-N	2.10	0.5 ± 0.1	0.01 ± 0.01	0.006 ± 0.000
Cu-P	2.25	1.1 ± 0.1	0.01 ± 0.01	0.006 ± 0.000
Cu-I	2.62	1.4 ±0.1	−0.06 ± 0.01	0.011 ± 0.001
Cu-Cu	3.03	1.6	0.22 ± 0.02	0.019 ± 0.003

The value for ΔE_0 extracted from this fit is −0.0 ± 0.5 eV.				
NHetPHOS complex **2** (film), R-factor = 0.003				
path	R / Å	N	ΔR / Å	σ^2 / Å^2
Cu-N	2.10	0.4 ± 0.2	0.01 ± 0.01	0.006 ± 0.001
Cu-P	2.25	1.3 ± 0.2	0.01 ± 0.01	0.006 ± 0.001
Cu-I	2.60	1.3 ± 0.1	−0.09 ± 0.01	0.009 ± 0.002
Cu-Cu	3.04	1.7	0.23 ± 0.03	0.018 ± 0.004

The value for ΔE_0 extracted from this fit is 0.5 ± 0.3 eV.				
NHetPHOS complex **3** (crystal), R-factor = 0.002				
path	R / Å	N	ΔR / Å	σ^2 / Å^2
Cu-N	2.10	0.4 ± 0.1	0.01 ± 0.01	0.006 ± 0.001
Cu-P	2.25	1.1 ± 0.1	0.01 ± 0.01	0.006 ± 0.001
Cu-I	2.63	1.4 ±0.1	−0.05 ± 0.01	0.011 ± 0.001
Cu-Cu	3.06	1.6	0.24 ± 0.02	0.025 ± 0.004

[*]For NHetPHOS complex **2**, single crystal x-ray diffraction revealed that Cu2 was disordered to an extent of approximately 6%. [28, 29] This disorder was neglected in the fits presented herein.[102]

In summary, the comparison of XANES and EXAFS spectra of crystalline and film samples indicates that the coordination geometry around the Cu(I) center as well as the type and distance of nearest neighboring atoms is retained upon solution processing of the investigated Cu(I) NHetPHOS complexes **1** and **2**.[102] "The aforementioned feature at 8995.3 eV in the XANES region in crystalline samples should be subject of further investigations."[102] This feature may be caused by minor angular distortions of the Cu_2I_2 unit. Even when the Cu-I and Cu-Cu-distances are retained (as indicated by the EXAFS results), there remains a certain degree of freedom regarding the angles between the Cu-Cu-I1 and Cu-Cu-I2 planes.[102] It is also confirmed that amorphous samples of complex **3** feature the expected coordination geometry.

To complement the Cu K XAS data solid-state ^{31}P NMR was used to compare the structures of the copper complexes in the various solid morphologies studied because of its high sensitivity: The method was used to study the alignment of membranes in biological samples.[119] "It has been shown that this method is also able to distinguish between two structural isomers of a Cu(I) complex in a crystalline matrix.[100, 101],,[102] Solid-state ^{31}P NMR CP/MAS [nuclear magnetic resonance (NMR) in cross polarization (CP) using magic angle spinning (MAS)] spectra of complex **1** prepared as a microcrystalline powder (Figure 32 a), as a ground-up neat film (Figure 32 b), and as an amorphous powder (Figure 32 e), were acquired. NHetPHOS complexes **2** and **3** were measured in the neat film state (Figure 32 c/d) and as amorphous powder (Figure 32 f/g).[102]

In solution, NMR peaks of NHetPHOS complexes **1**–**3** are broad and unstructured.[102] "As a consequence of the dynamic behavior mentioned in the previous section, ^{31}P NMR spectroscopy in $CDCl_3$ revealed only one P-component for the homoleptic complexes, despite the non-equivalence of the ligands.[88, 99, 120] The heteroleptic complexes feature at least two components, with one signal for the N,P-ligands and two broad, overlapping signals for the two P-donors.[28, 29],,[102] In solid state, this situation is quite different: NHetPHOS complexes **1** and **3** are dominated by a quartet line shape, as expected from the *J*-coupling between ^{31}P and the spin 3/2 nucleus ^{63}Cu or ^{65}Cu.[102] "As reported in earlier studies on phosphorus/copper complexes, the peak positions and intensities of the quartet deviate from a regular symmetric multiplet due to incompletely averaged residual dipolar coupling as a consequence of the $^{63}Cu/^{65}Cu$ quadrupolar interaction, and due the presence of the two isotopes of Cu.[100, 101, 121, 122],,[102] The spectra of NHetPHOS complex **2** (Figure 32 c/f) are composed of several overlapping quartets.[102] "In order to discern the different contributions, 2-dimensional ^{31}P-^{31}P correlation spectra were acquired for the amorphous powder samples of all complexes, which are shown below the corresponding 1-dimensional spectra (Figure 32 h-k)."[102] From the 2-dimensional spectra, one, three, and

two significantly differing *J*-coupling networks were identified for NHetPHOS complexes
1, **2**, and **3**, respectively.[102] This corresponds to one, three, and two cross-peaks in the 2-
dimensional spectra. In the spectra of NHetPHOS complex **3** as amorphous powder (Figure
32 d/g), the second signals are visible upfield of the major quartet.[102] "These minor
components also give rise to quartets, seen best in the 2-dimensional spectrum (Figure 32
k)."[102] These minor signals can be attributed to a negligible fraction of less than ~1% of
trace impurities, while the major fraction (> ~99%) of NHetPHOS complex **3** gives rise to
a single quartet.[102]

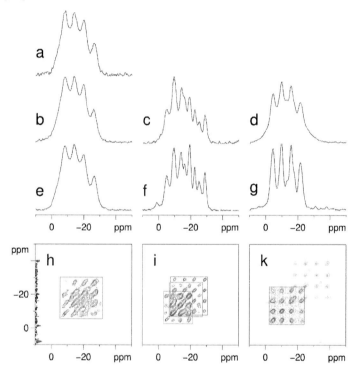

Figure 32: "Solid-state CP/MAS ^{31}P NMR spectra (a-g) and 2-dimensional ^{31}P–^{31}P correlation
spectra (h-k) of the Cu(I) NHetPHOS complexes **1** (a,b,e,h), **2** (c,f,i) and **3** (d,g,k), obtained at 242
MHz ^{31}P resonance frequency (600 MHz ^{1}H resonance frequency) under 25 kHz magic angle
spinning and using cross polarization from ^{1}H. Single crystals (a), thin films made by drop casting
(b-d), and amorphous powders (e-g) were measured. The spectra show one or several overlapping
distorted quartets of the different ^{31}P sites, coupled to ^{63}Cu or ^{65}Cu. The 2-dimensional ^{31}P–^{31}P
correlation spectra (h-k) of the amorphous powder samples are shown below the corresponding 1-
dimensional spectra. The *J*-coupling networks are marked in color."[102]†

NHetPHOS complexes 1–3 seem to show a different number of ^{31}P-resonances, despite their similar conformation.[102] "In principle, one would expect three separated signals, one for the bridging phosphine ligand, and two more for the monodentate phosphines. However, none of the samples showed this behavior; instead, signals are obviously overlapping."[102] This situation is quite similar to the results found in solution for NHetPHOS complexes,[28, 29, 48, 88] where the use of monodentate aryl phosphines led to one broad ^{31}P resonance in NHetPHOS complex 1, and two ill-resolved signals in NHetPHOS complexes 2 and 3, whereas the use of monodentate alkyl phosphines lead to two distinguishable signals (bridging aryl phosphine and monodentate alkyl phosphine).[102] In solid-state NMR, only NHetPHOS complex 2 showed three readily identifiable signals, which would match the three different phosphorus atoms of the bridging and the two monodentate phosphines.[102] The observation that NHetPHOS complexes 1 and 3 possess overlapping quartets (indicated by the non-concentric shape of the signals in the 2D plot) while several signals can be distinguished in complex 2, may be attributed to several effects, which are unlike the effects found in solution.[102] The differences between the complexes in solid state probably reflect the impact of different substituents on the chemical environment, as well as differences in their molecular structures, with NHetPHOS complex 2 deviating more from NHetPHOS complexes 1 and 3.[102] For example, as shown in Table 5, the Cu–Cu distance in 2 is increased and thus differs from those of NHetPHOS complex 1 and 3.[102] The angle between P–Cu–P for NHetPHOS complex 2 is the largest amongst the three studied compounds.[102] "In solution, the broad, unstructured signals may be related to an exchange of coordinated and dissociated ligands. Its sensitivity to small influences makes the solid-state ^{31}P-NMR spectra a useful tool to probe for small structural differences. A noticeable difference in the line width and relative intensities was observed for the different sample morphologies."[102] Comparing the respective spectra of the neat film and amorphous powder of compound 2 and 3, the signals of the neat film spectra exhibit a larger broadening. In complex 1, this increase in linewidth is not observed, only the microcrystalline sample seems to possess slightly sharper lines than the other two morphologies of NHetPHOS complex 1.[102] The origin of the differences in linewidth in NHetPHOS complex 2 and 3 could be either a different structural heterogeneity of the sample giving rise to dispersion in chemical shifts and couplings, a composition of micro-particles of different size giving rise to different local susceptibilities, or different transverse relaxation times.[102] "The fact that the multiplets seem to be composed of lines with rather symmetric line shapes would favor relaxation to be the dominant cause of line broadening. It would also be conceivable that the bulk powder preparations possess a higher density and rigidity than the film preparations. The larger extent of slow molecular motions in the case of the film material would then lead to increased transverse relaxation and linewidth."[102]

Comparing the ^{31}P NMR spectra obtained by differently prepared samples shows that the peak positions are the same for the respective crystalline (Figure 32a) and amorphous powders (Figure 32b–d) or thin films (Figure 32e–g) of each complex.[102] "Hence the different processing does not seem to drastically change the local structure around the phosphorus atoms."[102]

"Combining XANES and EXAFS results with further evidence from elemental analysis and solid state NMR leaves the structure where Cu is coordinated in average by 0.5 N, 1.5 P, 2.0 I, and 1.0 Cu atoms as the only chemically reasonable model structure."[102] Furthermore, these results show that the atomic arrangement within the probed molecules is comparable for single crystals and amorphous thin films.[102] This, in turn, suggests that the large color shift in combination with the preservation of the PLQY, and the emission decay times cannot be explained by dissociation of the complexes.[102] The theory of rigidochromic effects on quantum efficiency, which suggests both color and efficiency are influenced by matrix effects on the structure, does not seem to provide a sufficient explanation for the reported results.[102] "However, upon close examination minor differences are apparent in the XANES region of the XAS spectra and the line width in the NMR. The line width effects found in MAS ^{31}P-NMR suggest that the rigidity in bulk samples is indeed higher than in thin films. Surprisingly, this does not seem to influence the PLQY in this case. While the exact reason for this requires further investigation, one possible explanation could be a combination of rigidochromic effects with a distortion of the ground state geometry of the complexes. This could also be the reason for the differences found in the XANES spectra of the complexes."[102] According to DFT calculations the HOMO is located on the Cu_2I_2 unit of the NHetPHOS complexes.[29] "Changes in this unit, for example, by widening or flattening of the I-Cu-I angles, are thus likely to cause color shifts, while still being in accordance with all other spectroscopic results."[102] In summary, all results indicate that – within the accuracy of the used methods – NHetPHOS complexes **1–3** retain their principle structure even when preparing thin films by solution processing methods.[102] Both the chemical environment of the P-atoms, as well as the number, type, and distance of any ligands relative to the copper atoms are maintained. This suggests that the here-chosen complexes offer the opportunity to be used as emitters in optoelectronic devices even when processed as thin films from solution.[102]

4.2.2.2 Molecular structure of a NHetPHOS complex containing a bridging bis(diphenyl phosphine) ligand in an amorphous powder sample[‡]

As ligand exchange reactions and dissociation in solution, and distortion of the molecular structure in the excited state[123] are two mayor processes impairing the performance of Cu(I) complexes in optoelectronic devices, a NHetPHOS complex containing a bridging bis(diphenyl phosphine) ligand was developed. The molecular structure of NHetPHOS complex **4** together with two other members of the same complex family (NHetPHOS complexes **1** and **2**) is displayed in Figure 33. It can be assumed that ligand exchange reactions and dissociation in solution of NHetPHOS complex **4** containing an additional bridging ligand is less due likely due to the chelate effect and that the distortion upon excitation is reduced through steric stiffening of the structure.[60] The development of NHetPHOS complex **4** and its chemical characterization, photophysical measurements, and the respective analysis were performed by Dr. Daniel Volz and the OLED fabrication and specification was carried out by Dr. Ying Chen, Dr. Rui Liu, and Prof. Dr. Franky So.

[‡] The data and analysis reported in this chapter was previously published in the context of this work. Figures, tables, and wording is reprinted and adapted where appropriate with permission from ([81]   D. Volz, Y. Chen, M. Wallesch, R. Liu, C. Fléchon, D. M. Zink, J. Friedrichs, H. Flügge, R. Steininger, J. Göttlicher, C. Heske, L. Weinhardt, S. Bräse, F. So, T. Baumann, *Adv. Mater.* **2015**, *17*, 2538–2543.) Copyright (2015) John Wiley and Sons. The development of NHetPHOS complex **4** and its chemical characterization, photophysical measurements, and the respective analysis were performed by Dr. Daniel Volz and the OLED fabrication and specification was carried out by Dr. Ying Chen, Dr. Rui Liu, and Prof. Dr. Franky So.

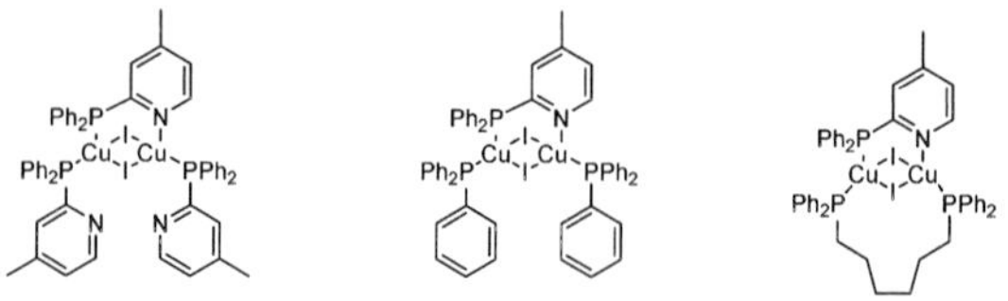

| NHetPHOS complex **1** | NHetPHOS complex **2** | NHetPHOS complex **4** |
| homoleptic | heteroleptic | heteroleptic |

Figure 33: Molecular structures of NHetPHOS complexes **1**, **2**, and **4**. NHetPHOS complexes **1** and **2** serve as reference for the interpretation of the Cu K XAS data.[60]‡

NHetPHOS complex **4** shows a high thermal stability (T_{decomp} = 290 °C according to thermogravimetric analysis, TGA).[102] When doping thin films of a host material (e.g., PYD2) with NHetPHOS complex **4**, a high PLQY (close to unity) is noted comparable to homoleptic and non-bridged heteroleptic complexes.[60] Compared to NHetPHOS complexes **1** and **2**, NHetPHOS complex **4** exhibits improved film formation properties and a reduced crystallization tendency, even when prepared as neat thin films without a host.[60]

The structural integrity of the molecule is of significant importance for its optoelectronic properties. However, single crystal x-ray diffraction as commonly deployed method for structure determination cannot be applied due to the low crystallization tendency of NHetPHOS complex **4** and the lack of suitable single crystals.[60] Instead, Cu K XAS was used to investigate the molecular structure of NHetPHOS complex **4** in an amorphous powder sample.[60] Based on these results, a first optimized OLED containing NHetPHOS complex **4** as emitter was fabricated. "This device marks two important milestones: First, the performance gap to triplet harvesting iridium emitter is closed by demonstrating that very high efficiencies can also be realized using Cu emitters."[60] Second, (to the best of our knowledge) a new quantum efficiency record for both solution- and vacuum-processed organic light emitting diodes with Cu(I) complexes as emitters is reached.[60]

As the atomic structure of NHetPHOS complexes **1** and **2** could be studied with single crystal x-ray diffraction due to their high crystallization tendency, these complexes serve as references for NHetPHOS complex **4**.[60] The qualitative analysis of the XANES region provides information concerning the local geometrical structure and the oxidation state of the central atom. The XANES spectrum for an amorphous powder sample of NHetPHOS complex **4** and crystalline samples of NHetPHOS complexes **1** and **2** (see Figure 33) are compared in Figure 34 to examine whether NHetPHOS complex **4** also features a distorted tetrahedral coordination geometry and whether the oxidation state of the Cu center is also +1.[60]

Cu(I) complexes exhibit characteristic low energy peaks in the region between 8983.0 eV and 8986.0 eV and one peak at 8990.0 eV.[112] "The former peaks have been assigned to $1s{\rightarrow}4p_x$ and $1s{\rightarrow}4p_{y,z}$ transitions.[112],[60] For NHetPHOS complex **4**, these characteristic peaks are observed at 8984.0 eV and 8990.5 eV, as for NHetPHOS complexes **1** and **2**.[60]

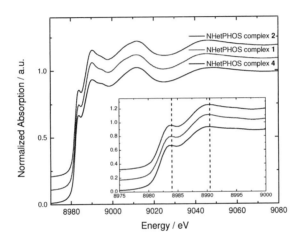

Figure 34: Cu K XANES spectra for an amorphous powder sample of NHetPHOS complex **4** and crystalline samples of NHetPHOS complexes **1** and **2**. Spectra are calibrated in energy and normalized to an edge jump of 1.0. The inset shows the XANES edge region in more detail. Vertical lines at 8984.0 and 8990.5 eV mark prominent features in the spectra. Compare chapter 4.2.2.1 for detailed discussion of the spectra of NHetPHOS complexes **1** and **2** (see Figure 33 for structure).[60]‡

"According to the ligand field splitting, the coordination number of Cu(I) and the symmetry of the ligand field around the central atom influence the energy position of the $1s{\rightarrow}4p$ transitions in Cu(I) complexes.[104, 106, 113, 124] For other Cu(I) iodide complexes, also with a distorted tetrahedral geometry around the Cu(I) ion, characteristic peaks at similar energies (8984.0 and 8990.5 eV) are observed.[106] Furthermore, information on the oxidation state of the Cu ion can be drawn from the XANES region: For Cu(II) complexes, intense peaks in the region between 8986.0 eV and 8988.0 eV are assigned to 1s→4p transitions and a pre-edge peak at 8979.0 eV corresponds to 1s→3d transitions."[60] Neither NHetPHOS complex **4** nor the homoleptic and heteroleptic reference complexes exhibit these features.[60] "This suggests that the d orbitals are fully occupied and that the oxidation state of copper is indeed +1.[112],[60] The overall similarity of the XANES spectra of NHetPHOS complex **4** and the NHetPHOS complexes **1** and **2** and the good agreement of the energies of the characteristic transitions (8984.0 eV and 8990.5 eV) with another Cu(I) iodide complex with a similar distorted tetrahedral geometry[106] indicate that the coordination geometry of NHetPHOS complex **4** is comparable and that the oxidation state is indeed +1.[60]

"Nearest-neighbor distances and coordination numbers can be derived from a quantitative analysis of the EXAFS region."[60] As the starting model for the fit, interatomic distances and coordination numbers derived from the single crystal x-ray diffraction analysis of the crystalline samples of NHetPHOS complexes **1** and **2** were used.[60] One Cu-N, Cu-P, Cu-I, and Cu-Cu single scattering path had to be included to obtain the best fit results for the amorphous powder sample of NHetPHOS complex **4** (shown in Figure 35).[60] Table 7 displays the obtained structural data. The Cu-N, Cu-P, and Cu-I scattering paths contribute to the feature between 1.0 and 2.1 Å in the Fourier-transformed $k^3\chi(k)$ (FT[$k^3\chi(k)$]) signal.[60]

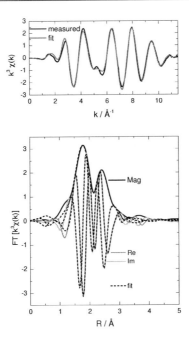

Figure 35: top: EXAFS data $k^3\chi(k)$, as measured (solid) and fit (dashed); bottom: FT[$k^3\chi(k)$] analysis - magnitude (Mag), real (Re) and imaginary (Im) parts, and their fits (dashed) for an amorphous powder sample of complex **1** (FT range $k = 2.2–10.5$ Å^{-1}).[60]‡

The Cu-I single scattering path has a complex shape and contributes additionally to the feature between 2.1 and 3.4 Å, together with the Cu-Cu path (see Figure 36).[60]

Table 7: EXAFS results for amorphous powder samples of NHetPHOS complex **4**.[60] "From left to right: single scattering paths used in the fit (path), interatomic distances (R), coordination numbers (N), deviation from the distances obtained from the XRD analyses (ΔR), and mean squared atomic displacement/Debye-Waller factors (σ^2). Note that the coordination number of Cu was restrained such that N(Cu-I) + N(Cu-Cu) = 3. The value for ΔE_0 extracted from this fit is -0.2 ± 0.7 eV, and the R-factor is 0.006."[60]‡

NHetPHOS complex **4** (amorphous powder), R-factor = 0.006			
path	R / Å	N	ΔR / Å	σ^2 / Å^2
Cu-N	2.08	0.5 ± 0.2	0.01 ± 0.01	0.007 ± 0.001
Cu-P	2.24	1.2 ± 0.2	0.01 ± 0.01	0.007 ± 0.001
Cu-I	2.59	1.8 ± 0.3	−0.05 ± 0.02	0.013 ± 0.003
Cu-Cu	3.03	1.2	0.28 ± 0.04	0.015 ± 0.005

For NHetPHOS complex **4** the value for the energy threshold offset ΔE_0 obtained from the fit is -0.2 ± 0.7 eV.[60] "The Debye-Waller factors σ^2 derived from the model fit range from 0.007 to 0.015 Å^2, with increasing values corresponding to longer interatomic distances."[60]

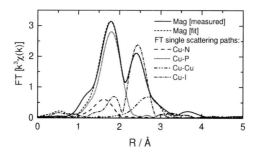

Figure 36: FT[$k^3\chi(k)$] data of crystalline sample of complex **1**: magnitude of the FT[$k^3\chi(k)$], best fit and single scattering paths used in the fit.[60]‡

Considering the intrinsically high correlation between coordination numbers and Debye-Waller factors, the coordination numbers derived from the fit of NHetPHOS complex **4** are in reasonable agreement with the expected mean coordination numbers (N) of the two Cu centers in a NHetPHOS complex ($N(N) = 0.5$, $N(P) = 1.5$, $N(I) = 2.0$ and $N(Cu) = 1.0$).[60] "The interatomic distances Cu-N, Cu-P, and Cu-I are in good agreement with the values found for the model compounds by single crystal diffraction, whereas the Cu-Cu distance obtained from the EXAFS analysis is about 0.28 Å longer. A similar difference between the Cu-Cu distances obtained by XRD and EXAFS for identical molecules was reported previously,[99] and might be due to the difference in measurement temperature (room temperature for the EXAFS measurements, 123 K for the XRD measurements), the different characters of EXAFS (sensitive to the short-range atomic order around the absorbing atom) and single crystal diffraction (long-range order), and/or a different effect."[60]

To sum up the analysis of the XANES and EXAFS region of the Cu K XAS data, it can be concluded the oxidation state of the Cu ions in NHetPHOS complex **4** is +1, that the coordination geometry around the Cu ions is similar to the distorted tetrahedral geometry in NHetPHOS complexes **1** and **2**, and distances and coordination numbers of the nearest neighbors of the Cu(I) ion are in good agreement with the expected values for a Cu(I) NHetPHOS complex.[60]

Note that the sample for XAS measurements was prepared from CuI with 98% purity, and hence a contribution from the Zn K-edge could also be observed in the EXAFS spectrum (Figure 37).[60] "Consequently, the k range for the Fourier Transformation (FT) was limited to $k = 2.3 - 10.5$ Å^{-1} to avoid interference of the Zn K edge. The samples for the herein reported photophysical measurements and OLED manufacture were prepared subsequently, using CuI with 99.999% (trace metals basis) purity."[60]

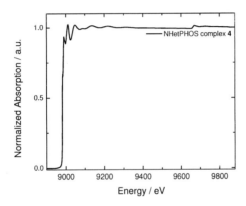

Figure 37: Cu K XAS spectrum for an amorphous powder sample of NHetPHOS complex **4**.[60]‡

NHetPHOS complex **4** promises to be an excellent emitter in a Cu(I)-based OLED device due to the well-defined structure, which was determined by XAS at the Cu K edge, and its good processability, low triplet energy, and high PLQY in PYD2-doped films.[60] Therefore, solution-processed multilayer OLEDs with NHetPHOS complex **4** were fabricated. The complete device configuration of these OLEDs is depicted in Figure 38, left: "A 30 nm layer of PEDOT:PSS was spin-coated on the ITO substrate, followed by 45 nm of PLEXCORE UT-314, which was spin-coated and cross-linked as described in a recent publication.[125] The EML, consisting of NHetPHOS complex **4** doped in PYD2 (27 nm) was then spin-coated on top of the HTL and dried on a hot plate. A 50 nm film of 3TPYMB ETL was deposited by thermal evaporation, followed by 2 nm LiF and an Al cathode of 100 nm thickness."[60]

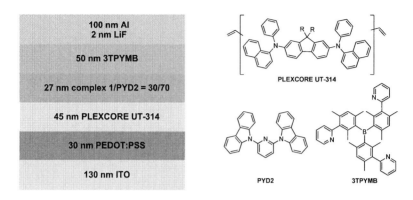

Figure 38: "Device architecture and molecular structures of the organic materials used herein."[60]‡

"The turn-on voltage of the resulting OLED device was found at 2.6 ± 0.1 V, while a maximum brightness of 10.000 ± 1000 cd m^{-2} was achieved at 10 V."[60] Figure 39 shows the current efficiency as a function of luminance of the optimized device (top), together with its current density and luminance as a function of voltage (bottom).[60] "The current efficiency is 71 ± 2 cd A^{-1} at 100 cd m^{-2} and 47 ± 2 cd A^{-1} at 1000 cd m^{-2}. The peak current efficiency is 73 ± 2 cd A^{-1} at ca. 40 cd m^{-2}, corresponding to a peak external quantum efficiency (EQE) of $23 \pm 1\%$."[60]

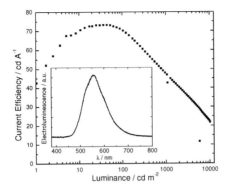

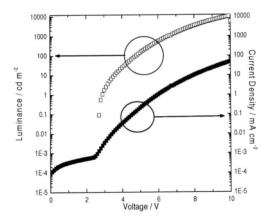

Figure 39: "Top: Current efficiency vs. luminance; Inset: electroluminescence spectrum of the device. Bottom: Luminance (left ordinate) and current density (right ordinate) versus bias voltage.[60]‡

"To the best of our knowledge, this is the highest EQE value ever reported for OLEDs with solution-processed emissive layer based on Cu(I) emitters and is comparable to the efficiency of state-of-the-art thermally evaporated devices with Ir(III) emitters. The luminous efficacy of the OLED device presented here is significantly higher than those of current solution-processed record devices with green Cu emitters (49.5 cd/A)[126]and even the best vacuum-processed OLEDs with Cu emitters (65.3 cd/A).[127]"[60]

The EQE of an OLED may be calculated by using the following, simplified equation:[25]

$$EQE = \chi_{out} \times \Phi_{PL} \times \beta \times \gamma \tag{4.4}$$

The internal quantum efficiency (IQE) is given by:

$$IQE = \Phi_{PL} \times \beta \times \gamma \tag{4.5}$$

χ_{out} represents the optical outcoupling factor, Φ_{PL} the PLQY of the emission layer (determined to be 92 ± 5%), β the fraction of excitons that may be harvested, and γ the charge balance factor. "χ_{out} accounts for loss processes due to waveguiding and interaction with the metal electrode, and may be modelled if the refractive indices of all layers are known. Assuming that the device is a Lambertian light source with randomly oriented emitters, and that the refractive indices are similar to other OLED materials (1.7 to 1.9), a common approximate value of $\chi_{out} = 0.25 \pm 0.05$ can be assumed.[25, 128] For the singlet harvesting mechanism, both singlet and triplet excitons may be harvested, meaning that β equals unity. The charge balance, e.g., the ratio between holes and electrons in the emissive layer, often depends on the current. One could speculate that the charge carriers are balanced in the peak region of the luminance-efficiency plot (Figure 39), therefore – $\gamma \approx 1$."[60]

To derive an estimate for the internal quantum efficiency (IQE), the highest experimental value for EQE (23%) and $\chi_{out} = 0.25 \pm 0.05$ were used. "If all aforementioned assumptions are valid, the internal quantum efficiency IQE lies between 77 and 100%, with the most likely value being 92%, close to unity. This finding is corroborated by estimating the IQE using the previously discussed assumptions for β, which yields an IQE identical to the PLQY, 92 (± 5)%, as derived above."[60]

"Measurement and optimization of device lifetime was not within the scope of this study. So far, the highest lifetime ever reported for a Cu emitter is 440 hours at 100 cd m^{-2} for a non-optimized vacuum processed device by Thompson and coworkers.[25] Optimizing the lifetime of solution-processed devices is considerably more time-consuming than for vacuum processes, because variation of charge transport layers requires the development of crosslinkable materials and careful control of layer morphology. A common strategy is to use vacuum-processable model emitters and realize the resulting, optimized stack with crosslinkable analogues.[129],[60] This way, the few issues concerning the OLED device can be solved: "Figure 39 shows a decreasing efficiency at high luminance, a phenomenon commonly described as "efficiency roll-off".[130] While roll-off is often encountered in non-optimized OLEDs, the roll-off in this device is more pronounced than in highly-optimized evaporated phosphorescent OLEDs. The loss of charge balance at high injection could be responsible for the severe efficiency roll-off in some phosphorescent OLEDs.[130] In fact, the sharp increase in the efficiency in the reported device at low luminance is a typical sign of a lack of charge balance at low injection conditions."[60]

The structure determination of NHetPHOS complex **4** by means of XAS at the Cu K edge enabled the deployment of this material as emitter in OLEDs. Despite the given potential for optimization of the OLED architecture, the solution-processed OLED based on NHetPHOS complex **4** as emitter, finally bridges the efficiency gap between Cu emitters and attain a level that is comparable with today's best green phosphorescent OLEDs using iridium or platinum. To further improve the OLED performance, a better understanding of the NHetPHOS complexes will be established and the electronic structure of NHetPHOS complexes will be studied by XES and RIXS at the N K edge.

4.3   Electronic interaction of the Cu atom and the ligands

4.3.1 Methods: X-ray emission spectroscopy and resonant inelastic soft x-ray scattering at the N K edge

In non-resonant x-ray emission spectroscopy (XES), a sample is irradiated with x-ray photons well above the absorption edge, some of the x-ray photons are absorbed by the atoms in the sample and core electrons are ejected. The created core hole can be filled by an electron from a higher-lying level. In this process, energy is released. This energy can either be released in form of an x-ray photon (i.e., x-ray fluorescence) or transferred to another electron, which is then ejected from the atom (i.e., in the Auger process). In the range of soft x-rays, the relative yield for x-ray fluorescence is much lower (<< 1%) than for the Auger process. Therefore, XES requires high excitation intensities and efficient detection.

The intensity of a non-resonant x-ray emission process is described by Fermi`s Golden Rule in equation (4.1) and (4.3), replacing $hv_{absorption}$ by $hv_{emission}$.[109]

$$I(hv_{emission}) \propto \left| \langle \Psi_f | H_s | \Psi_i \rangle \right|^2 \rho_{lp}(E_i) \; with \; (E_i = E_f + hv_{emission}) \qquad (4.6)$$

The dipole selection rules specified for x-ray absorption in 4.2.1.2 also apply for the x-ray emission process. XES allows the investigation of the local electronic environment of a specific element in a sample. In contrast to PES, XES offers an extended information depth and probes the bulk structure of a sample. Furthermore, non-conducting samples can be studied, since XES, as a photon-in/photon-out process, is not affected by charging.

In resonant inelastic soft x-ray scattering (RIXS), a sample is excited with x-ray photons in the range of the absorption edge of the studied element. In this case, the probability of the transition from the initial state $|\Psi_i\rangle$ with energy E_i to the intermediate state $\langle \Psi_m|$ with energy E_m upon absorption of a photon $hv_{absorption}$, and the transition from the intermediate state to the final state $\langle \Psi_f|$ with energy E_f upon emission of photon $hv_{emission}$ has to be described as a one-step process. This can be done by means of second-order perturbation theory.[131, 132] According to this, the cross section for the resonant absorption of a photon hv_{ab} and the emission of a photon hv_{em} within the solid angle Ω is given by the Kramers-Heisenberg dispersion formula:[132]

$$\frac{d^2\sigma(v_{ab})}{dv_{em}d\Omega} \propto \sum_e \left| \sum_m \frac{\langle \Psi_f | \hat{A}_{em}\hat{p} | \Psi_m \rangle \langle \Psi_m | \hat{A}_{ab}\hat{p} | \Psi_i \rangle}{E_m - E_i - hv_{ab} - i\Gamma_m/2} \right|^2 \delta(hv_{ab} - hv_{em} + E_i - E_m) \quad (4.7)$$

Γ_m is the lifetime broadening of the intermediate state $|\Psi_m\rangle$. The denominator sets a requirement for resonant absorption as intermediates states with energies $E_m = E_i + hv_{ab}$ are preferred.

RIXS is a valuable tool for the investigation of the electronic structure of solid,[133, 134] liquid,[135-137] and gas phase[135, 138] samples and gives detailed information about the local chemical environment of specific elements. A series of emission spectra – one for each excitation energy of a regular x-ray absorption scan – can be presented in a two-dimensional RIXS map.[134, 139] In a RIXS map, the emission intensity is color-coded and given as a function of emission energy and excitation energy. This approach allows the identification of separated emission and absorption features and therefore, provides more detailed information than sole emission or absorption spectroscopy.

The use of XES and RIXS seems advantageous for the investigation of the electronic structure of luminescent metal organic complexes. Whereas photoelectron spectroscopy mainly probes the surface of a sample, XES and RIXS offer an extended information depth and probe the bulk structure. This is especially favorable for materials that can be processed from solution like Cu(I) complexes, as surface contaminations cannot be completely excluded in this process. As XES and RIXS further offer site-specific information, the N K edge was chosen herein, to study two archetypal members of the Cu(I) NHetPHOS complex family as well as the complementary N-aromatic ligand. The nitrogen atoms serve as indicators for the structural differences in these materials. Therefore, the emphasis of this study is directed towards occupied and unoccupied molecular orbitals formed upon coordination of the Cu and N atom, which are expected to influence the luminescence properties of the Cu(I) NHetPHOS complexes.

4.3.2 Result: X-ray emission spectroscopy and resonant inelastic soft x-ray scattering at the N K edge[§]

4.3.2.1 Study of two Cu(I) NHetPHOS complexes and the complementary ligand

In general, ligand-centered transitions, inter-ligand transitions, d-s transitions on the metal centers, as well as charge transfer transitions between the metal center and ligands, entail the luminescence of Cu(I) complexes. For Cu(I) NHetPHOS complexes, it was assumed that metal-to-ligand charge-transfer (MLCT), ligand-to-ligand charge-transfer (LLCT), and/or metal-halide-to-ligand charge-transfer (M+X)LCT transitions are involved in the emission process, as the emission color of Cu(I) NHetPHOS complexes can be tuned by systematical variation of the ligand system.[48, 50, 51] However, the electronic structure of NHetPHOS complexes has not been studied experimentally in detail so far.

A better general understanding of Cu(I) NHetPHOS complexes and a greater insight into the electronic interaction between the Cu(I) atom and the ligands is required, to further improve the design of these materials and thereafter yield solution-processed OLEDs with longer device lifetimes.

The Cu(I) NHetPHOS complexes consist of a butterfly shaped Cu_2I_2-core with a bridging N-aromatic ligand and two terminal ligands, as depicted in Figure 40.[26] Ligand **3** consists of a methyl-pyridine unit, which is connected to a diphenylphosphine moiety. Pyridine and its derivatives, which have been studied with x-ray spectroscopic techniques previously, can serve as reference due to the similarity of their molecular structure around the N-atom to that in ligand **3**. The geometric structure around the N-atom is similar for NHetPHOS complexes **1** and **2**. As no crystals suitable for single crystal x-ray diffraction measurements could be obtained for the pristine ligand **3**, known bond lengths and angles for ligand **4**[140] are displayed, as the molecular structure is identical with that of ligand **3** besides of the missing methyl substituent on the pyridine ring. For comparison, the bond lengths and angles of the investigated molecules are summarized in Table 8.

[§] Theoretical calculations were conducted by Fernando D. Vila (PhD) from the group of Prof. John J. Rehr (PhD) at the University of Washington. XES and RIXS measurements at the ALS were performed in cooperation with Andreas Benkert, Dirk Hauschild, Dr. Dagmar Kreikemeyer-Lorenzo, Frank Meyer, Dr. Monika Blum, Prof. Dr. Marcus Bär, Dr. Regan G. Wilks, Dr. Wanli Yang, Prof. Dr. Clemens Heske, and Dr. Lothar Weinhardt.

Table 8: Selected bond lengths and angles for NHetPHOS complexes **1** and **2** and ligand **4**.*

	N-C1	N-C5	C1-N-C5	N-C5-P	Cu-N
		NHetPHOS complex **1**			
N binding to Cu	1.330(8)	1.334(8)	117.5(5)	117.2(4)	2.078(5)
N not binding to Cu	1.333(10)	1.342(9)	114.9(7)	116.5(5)	-
	1.321(9)	1.332(8)	116.3(6)	113.8(5)	-
		NHetPHOS complex **2**			
	1.355(3)	1.343(3)	116.3(2)	116.38(19)	2.090(2)
		ligand **4**			
	1.342(3)	1.324(5)	116.8(3)	111.9(2)	-

* Estimated standard deviations refer to the last digit printed. Bond lengths and angles for ligand **4** were taken from [140].

In the homoleptic NHetPHOS complex **1**, all three organic ligands are the same, whereas in the heteroleptic NHetPHOS complex **2** the two terminal ligands are triphenylphosphine. Therefore, NHetPHOS complex **1** contains three non-equivalent N-atoms: One, which is binding to one of the Cu(I)-atoms and two, which are non-binding. In NHetPHOS complex **2**, there is one N-atom, which is binding to one of the Cu(I)-atoms.

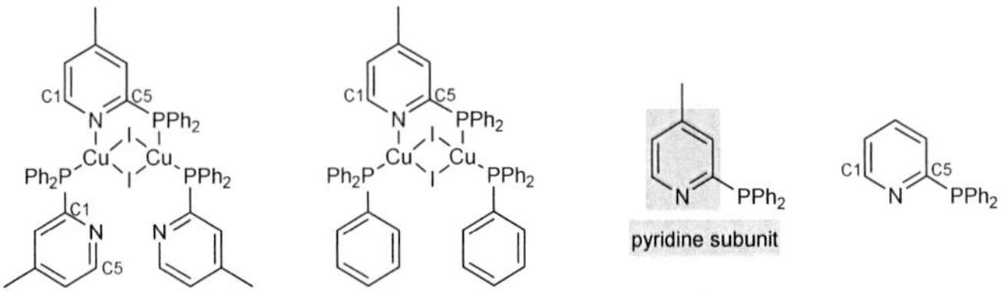

homoleptic NHetPHOS complex **1**

heteroleptic NHetPHOS complex **2**

ligand **3**

ligand **4**

N-atom binding to Cu-atom
N-atom not binding to Cu-atom

Figure 40: Structural formulas of the homoleptic and heteroleptic Cu(I) NHetPHOS complexes **1** and **2** and the N-aromatic ligand **3**.

Due to these differences in the coordination environment, the N-atoms serve as indicators for the differences in the molecular and related electronic structure of NHetPHOS complexes **1** and **2** and ligand **3**. DFT calculations on these materials predict that the LUMO is mainly located on the N-aromatic ligand and the HOMO mainly on the Cu_2I_2-

core with the nitrogen linking both components. Therefore, the N K edge was chosen for this comparative RIXS and XES study.

4.3.2.2 N K RIXS maps

Figure 41 shows the N K RIXS maps of NHetPHOS complexes **1** and **2** and the ligand **3**.

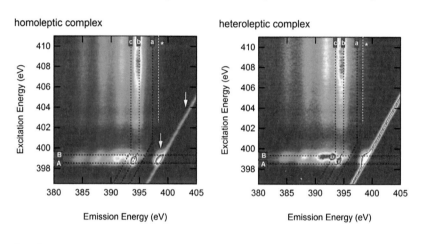

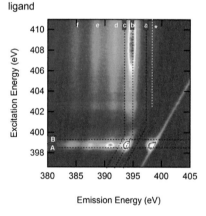

Figure 41: N K RIXS maps of NHetPHOS complexes **1** and **2** and ligand **3**. The x-ray emission intensity is color-coded and shown as a function of the excitation and emission energies. The maps are normalized in emission intensity to the maximum of feature b in the non-resonant range. Dashed red lines, together with red bars labeled A, B, and a, b, c, indicate the summation windows for the resonant XES and decay-channel-specific XAS spectra.[§]

The RIXS maps can be divided into two areas: In the excitation energy range below ~ 400.0 eV, the emission features shift parallel to the Rayleigh line, indicating inelastic scattering with specific loss energies. This energy loss can be described as electronic Raman scattering. In the range above an excitation energy of ~ 400.0 eV, these emission features are present as vertical lines. In all three maps, the Rayleigh line (white arrow in Figure 41, top left) with equal excitation and emission energies is visible as a diagonal.

A strong resonance of the Rayleigh line can be observed around 399.0 eV, which is attributed to a core-excitonic state, which gives rise to an enhanced participant decay of the excited electrons. The intense resonance extending to lower energies from the Rayleigh line (yellow arrow in Figure 41, enlargement in Figure 42) can be attributed to vibrational excitations in the final state. The shape of this resonance differs for each material as illustrated in Figure 42. While for ligand **3**, this resonance is dominated by a circular-shaped feature, centered at an excitation energy of 398.7 eV, for NHetPHOS complex **2**, this resonance is elongated and extends to higher excitation energies (with the maximum at an excitation energy of 399.0 eV). For NHetPHOS complex **1**, this resonance appears to be comprised of two components resembling the circular and elongated feature from ligand **3** and NHetPHOS complex **2**. This is in accordance with NHetPHOS complex **1** containing one binding N atom (as present in NHetPHOS complex **2**) and two non-binding N-atoms (as present in ligand **3**). As the investigated materials exhibit a multitude of closely spaced vibronic states in the final state, the vibronic fine structure is not resolved in this resonance, in contrast to previous studies at the N K edge, e.g., of aqueous ammonia.[137]

Overall, the most prominent emission and absorption features in the RIXS map of NHetPHOS complexes **1** and **2** resemble the ones of the ligand, as the molecular and geometric structure of the pristine ligand **3** and the ligand in NHetPHOS complexes **1** and **2** is similar.

Nevertheless, there are distinct differences in the RIXS maps of ligand **3** and NHetPHOS complexes **1** and **2**. First, the energy position and relative intensity of the absorption threshold deviates for the complexes and the ligand. For ligand **3**, all emission features appear at an excitation energy of approx. 398.1 eV, while for NHetPHOS complexes **1** and **2**, the energetic position of the absorption onset differs for different emission features. The maximum intensity of the absorption threshold lies between approximately 398.1 eV and 398.6 eV.

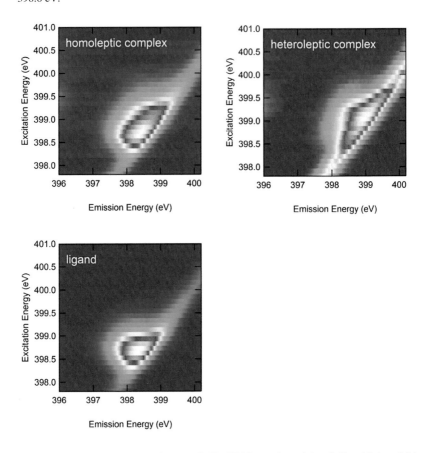

Figure 42: Detail in the N K RIXS maps of NHetPHOS complexes **1** (top left) and **2** (top right) and ligand **3** (bottom).[§]

Furthermore, there are differences in the relative intensity of the emission and absorption features of NHetPHOS complexes **1** and **2** and ligand **3**. In the following, a comparative analysis of these differences will be conducted on the basis of resonant and non-resonant x-ray emission and absorption spectra.

4.3.2.3   Resonant and non-resonant N K x-ray emission spectra

The analysis of the x-ray emission spectra provides information about the occupied MOs of the complexes and the ligand.

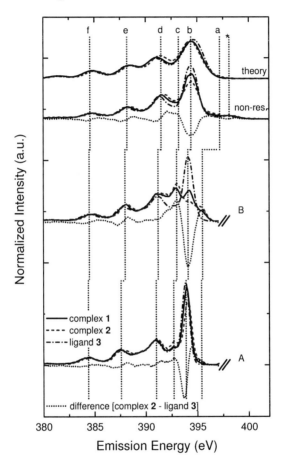

Figure 43: Resonantly excited N K XES spectra A and B, together with non-resonant excited (non-res.) spectra of NHetPHOS complexes **1** and **2** and ligand **3** as well as calculated spectra. All spectra were normalized to the area under the curves. The dotted spectra represent the difference of NHetPHOS complex **2** and ligand **3**. Spectra A and B were obtained by summation over the spectra with excitation energies between 398.2 and 398.8 eV, as well as between 399.0 and 399.6 eV in the RIXS maps.[§]

Resonant XES spectra were obtained from the RIXS maps by summing up the emission spectra for excitation energies between 398.2 and 398.8 eV, as well as between 399.0 and 399.6 eV (dashed red lines A and B in Figure 41 with bars indicating the summation windows) and are presented together with non-resonantly excited spectra in Figure 43. Distinct spectral changes between non-resonant and resonant spectra as well as between the three samples can be observed. Six distinct spectral features (a–f) can be identified in the non-resonant spectra of NHetPHOS complexes 1 and 2 and ligand 3, as illustrated in Figure 43.

The spectra can be interpreted by comparison with the literature: In a previous study of poly-pyridines and pyridine, Magnuson et al.[141] provided static exchange calculations to assign these features as follows: Feature b has been attributed to excitations from π electron states and the lone-pair n (non-bonding) orbital, which has σ symmetry ($1a_2$ in the poly-pyridines and $11a_1(n)$ in pyridine). Feature d is supposed to be due to excitations from both π and σ electronic states with dominating σ character, while peaks e and f are ascribed to excitations from σ electronic states.[141]

The non-resonant spectra for NHetPHOS complexes 1 and 2 and the ligand 3 all show the same four main spectra features (b, d, e, and f). However, features d, e, and f show a shift by approx. 0.2 eV for NHetPHOS complex 1 and 0.4 eV for NHetPHOS complex 2 to higher emission energies in comparison to ligand 3. Furthermore, the relative intensity of feature b and region c varies for each material. For NHetPHOS complex 1 (2), feature b drops to about 85% (70%) of the corresponding intensity in the spectrum of ligand 3. At closer inspection, an additional feature (labeled a in Figure 41 and Figure 43) at 397.2 eV can be identified, which can only be observed in the non-resonant spectra for NHetPHOS complexes 1 and 2.

In the resonant spectra A and B, features a–f are shifted to lower emission energy in comparison to the non-resonant spectra, which can be attributed to spectator shifts, which arise from the presence/absence of the excited electron in the resonant/non-resonant case and its interaction with the remaining electronic system. These shifts are indicated by the vertical dashed lines in Figure 43.

There are significant differences for the resonant spectra A and B of NHetPHOS complexes 1 and 2 and ligand 3. The most dominant feature of the resonant spectra A and

B of ligand **3** is feature b. For ligand **3**, Spectra A and B are very similar. Furthermore, spectra A for NHetPHOS complexes **1** and **2** are – besides slight variations in energy position (0.1 and 0.2 eV) and intensity of feature b – similar to spectrum A of ligand **3**. In contrast, spectra B for NHetPHOS complexes **1** and **2** differ strongly from that of ligand **3**. Compared to the spectrum of ligand **3**, the intensity of feature b is reduced for NHetPHOS complex **1** and even more for NHetPHOS complex **2**. Furthermore, feature a, which is visible as a small shoulder in the non-resonant spectra of the complexes, is strongly pronounced in the resonant XES spectra B but little pronounced in spectra A.

As we compare the pristine ligand **3** and NHetPHOS complex **2**, which contains only one N-atom binding to a Cu(I)-atoms, and NHetPHOS complex **1**, which contains one N-atom binding to a Cu(I)-atom and two which are non-binding, these differences can be explained by the formation of new occupied and unoccupied molecular orbitals upon formation of the Cu-N bond in the complex in comparison to the pristine ligand.

The following hypothesis can explain the differences between the resonant spectra A and B: For low excitation energies at the absorption threshold, the N 1s electron might be excited into the low-lying unoccupied MOs, centered at the N-atom. The emission might occur from an occupied orbital, centered at the N-atom. As the influence of the Cu-N bond is low on these N-centered orbital, spectra A for NHetPHOS complexes **1** and **2** and ligand **3** are similar. For increasing excitation energies, the N 1s electron could be excited into an unoccupied orbital, centered at the Cu_2I_2-core, for NHetPHOS complexes **1** and **2**. Due to the enhanced wave function overlap between the excited core level and the decaying valence orbitals of the Cu_2I_2-core in this case, the emission might occur from occupied orbitals, centered at the Cu_2I_2-core. Spectra B of the NHetPHOS complexes **1** and **2** might differ from spectra B of ligand **3**, as these spectra are mainly dominated by emission from occupied orbitals centered at the Cu_2I_2-core for the former and at the N-atom for the latter. As feature a is enhanced for higher excitations energies, the emission might be attributed to high-lying occupied MOs, centered at the Cu_2I_2-core. The relative change in the spectra of NHetPHOS complex **2** and the ligand **3** are even more apparent in the difference spectra of NHetPHOS complex **2** and ligand **3** displayed in Figure 43.

The theoretical simulated spectra, calculated using StoBe, are also displayed in Figure 43. The relative energy position and intensity of features a–f are well reproduced in the calculation. However, the variation in intensity of features b and shoulder c for the different materials is underestimated in the theory. For NHetPHOS complexes **1** and **2** as well as ligand **3**, there is a feature (marked with an asterisk in Figure 43) with low intensity at 398.1 eV. This feature is also visible in the N K RIXS maps above 402.5 eV in excitation energy. This feature is not reproduced in the calculation, which suggests that the origin of this feature is a core-excitonic emission.

It can be concluded that the chemical environment around the absorption site has a significant impact on the shape of the XES spectrum. Differences in the resonant and non-resonant emission spectra of the complexes and the ligand might be attributed to the formation of new unoccupied and occupied electronic molecular orbitals upon coordination of the N-atom to a Cu(I)-center in NHetPHOS complexes **1** and **2**. Comparison of the experimental and theoretical non-resonant XES spectra allowed the identification of the most significant emissive states and a core-excitonic state.

4.3.2.4   N K x-ray absorption spectra

To further analyze the RIXS maps, summations of the intensity over fixed emission energy windows (a, b, c, and PFY) in the RIXS maps – so-called decay-channel-specific (DCS) XAS spectra – of NHetPHOS complexes **1** and **2** and ligand **3** are presented in Figure 45. The emission energy windows are depicted in the maps in Figure 41 by red bars labeled a, b, and c, while spectrum PFY was generated by integrating the entire spectator region (380.0 eV–396.0 eV). In the Raman regime of the map, these emission energy windows were shifted parallel to the Rayleigh line, while they were at fixed emission energies above as shown in Figure 41 (dashed red lines a, b, and c with bars indicating the summation windows). For comparison, the N K inner shell electron energy loss spectrum (IEELS) of pyridine in the gas phase is shown as well (bottom spectrum).[142]

In literature, the most prominent features of poly-pyridines and pyridine have been assigned as follows: As the ground state of the pyridine molecule has C_{2v} group symmetry and its two lowest unoccupied MOs are of b_1 (LUMO) and a_2 (LUMO + 1) symmetry, the N_{1s} to LUMO ($3b_1$) transition at 398.8 eV is pronounced, whereas the N_{1s} to $2a_2$ transition lacks intensity for pyridine.[141] Due to a lower symmetry of the poly-pyridines, the N_{1s} to (LUMO + 1) transition is pronounced as a high energy shoulder (at 399.5 eV) of the N_{1s} to

LUMO transition.[141] These transitions could be attributed to feature 1 in Figure 45 for the NHetPHOS complexes **1** and **2** and ligand **3**. Feature 2 has been assigned to excitations to Rydberg or mixed valence/Rydberg orbitals.[143] Deviating assignments of feature 3, which is close to the ionization potential, can be found in the literature: A mixture of valence and Rydberg-type transitions[144] as well as an additional contribution of double excitations as described for the N_2 molecule[145-147] have been discussed.[143]

In the DCS-XAS spectra a, b, and c, the aforementioned deviation in energy position and relative intensity of the absorption threshold is visible. Ligand **3** does not exhibit distinct absorption features in spectrum a, whereas feature 1 is increasingly pronounced for NHetPHOS complexes **1** and **2**. In spectrum b, the intensity of feature 1 is reduced (65%) for NHetPHOS complex **1**, containing one N-atom binding to a Cu-atom and two N-atoms not binding to a Cu-atom, and even more for NHetPHOS complex **2** (40%), containing one N-atom binding to a Cu-atom. The energy position of this feature is in good agreement for the NHetPHOS complexes **1** and **2** and ligand **3** (while NHetPHOS complex **1** shows no shift, NHetPHOS complex **2** exhibits a slight shift of 0.1 eV to higher energies). For spectrum c, the trend in intensity of feature 1 is reversed with NHetPHOS complex **2**/ligand **3** exhibiting a higher/lower relative intensity than NHetPHOS complex **1**. The energy position of feature 1 is shifted by 0.4 eV to higher energies for NHetPHOS complexes **1** and **2** in comparison to the ligand **3** and is in good agreement with the energy position of feature 1 in the spectra a.

It can be speculated that the shifted feature is attributed to a transition into low-lying unoccupied MOs related to the Cu-N bond. As the pyridine-ligand is not highly symmetric, a change in symmetry upon coordination of the N- and Cu-atom is less likely the explanation for this shift than in the case of pyridine and poly-pyridines. Feature 2 and 3 are similar in shape and energy position in all spectra – except for spectrum a of ligand **3**. Spectra a, b, and c represent competing decay processes, which might correspond to an excitation into Cu_2X_2-centered, low-lying unoccupied MOs for spectra a and c and into N-centered, low-lying unoccupied MOs for spectrum b. The decay-channel-specific excitation (spectrum a) is only present for the NHetPHOS complexes **1/2** and not for ligand **3**, and might therefore be attributed to low-lying MOs that have been formed upon the Cu-N binding. The assignment of emission feature B to high-lying occupied MOs, associated with the Cu_2X_2-core, and absorption feature a to low-lying unoccupied MOs, associated with the Cu-N bond, for NHetPHOS complexes **1** and **2**, respectively, could provide experimental evidence, that the emissive transition is a metal-halide-to-ligand charge transfer, as previous DFT calculations indicated.

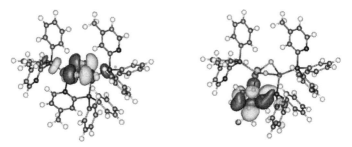

Figure 44: HOMO (left) and LUMO (right) frontier orbital plot of NHetPHOS complex **1** based on DFT calculations.[148]**

Frontier orbital plots of NHetPHOS complex **1**, derived by DFT calculations, are presented in Figure 44. The HOMO (left) is mainly centered on the Cu_2I_2 core, while the LUMO (right) is located on the N-aromatic ligand. According to this, upon photophysical excitation, an electron is excited from the HOMO into the LUMO via a metal-halide-to-ligand charge-transfer (M+X)LCT transitions. Radiative decay of the excited state leads to an optical emission.

** [148] D. M. Zink, D. Volz, L. Bergmann, M. Nieger, S. Bräse, H. Yersin, T. Baumann, *Proc. SPIE 8829* **2013**, 882907. Copyright 2013 Society of Photo Optical Instrumentation Engineers. One print or electronic copy may be made for personal use only. Systematic reproduction and distribution, duplication of any material in this paper for a fee or for commercial purposes, or modification of the content of the paper are prohibited. http://dx.doi.org/10.1117/12.2028619.

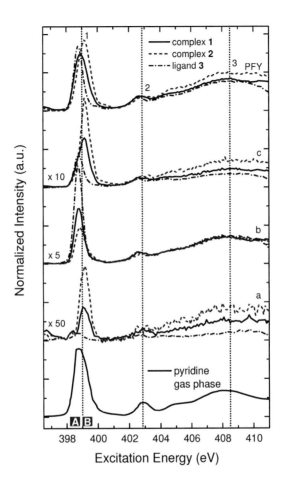

Figure 45: Bottom: N K ISEELS reference spectrum of pyridine in the gas phase, taken from [142]. Second from bottom to top: N K DCS XAS spectra a, b, c, and PFY of NHetPHOS complexes **1** and **2** and ligand **3** obtained by summation as depicted in Figure 41 (dashed red lines a, b, and c with bars indicating the summation windows). Region a – summation from 396.9 to 397.55 eV and from -3.5 to 2.9 eV; region b – summation from 394.5 to 395.1 eV and from -5.1 to 4.5 eV; region c – 393.1 to 393.7 eV and from -6.5 to 5.9 eV emission energy in the non-resonant range and loss energy in the resonant range, and PFY – summation from 380.0 to 396.0 eV emission energy. Prominent features 1, 2, and 3 at 399.0 eV, 403.0 eV, and 408.5 eV are marked with dotted lines. Red bars A, B indicate the summation windows for the resonant XES spectra. The intensities of the DCS XAS spectra correspond to the intensities in the particular map displayed in Figure 41.[§]

4.3.2.5 Summarizing discussion of the electronic interaction of the Cu atom and the ligands

Using XES and RIXS at the N K edge, two luminescent homo- and heteroleptic Cu(I) NHetPHOS complexes as well as the corresponding non-luminescent N-aromatic ligand were characterized. This approach could enable the assignment of absorption and emission features in the N K RIXS map to occupied and unoccupied molecular orbitals formed upon binding of the N- and Cu-atoms, which have either dominating N- or Cu_2I_2-character. In combination with theoretical calculations, we were able to identify the N K core-excitonic state of the Cu(I) complexes. Furthermore, our results might provide first experimental proof that the emissive transition is a metal-halide-to-ligand charge transfer, as previous DFT calculations indicated (Figure 44). Theoretical calculations, which allow the assignment of specific molecular orbitals to absorption and emission features, are still ongoing.

We will now review different methods to assess the HOMO and LUMO energies of the Cu(I) emitter, on the basis of a study of selected members of the Cu(I) NHetPHOS complex family, as these parameters are important for OLED fabrication-

4.4   Study of the HOMO and LUMO energies in selected Cu(I) complexes

4.4.1 Critical review of different methods for the assessment of the HOMO and LUMO energies

4.4.1.1 Optical spectroscopy

Common techniques for the assessment of the band gap and/or HOMO and LUMO energies, sometimes w.r.t. the vacuum level or the ionization potential and electron affinity, include optical spectroscopy, electrochemical methods like cyclic voltammetry, and photoelectron spectroscopy in air (PESA). In the context of this work, the aforementioned methods were compared to UPS and IPES measurements for a representative set of luminescent Cu(I) complexes. In the following, basics as well as advantages and limitations of each method will be discussed.

In literature, the band gap of a semiconductor is frequently derived from the low-energy onset of the optical absorption or diffuse reflectance spectrum of semiconductors.[149-151]

The dependence between the absorption coefficient and the excitation energy gives information about the combined density of states at the band edges. A typical curve representing this dependence is given in Figure 46.

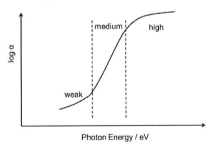

Figure 46: Typical curve representing the energy dependence of the logarithmic absorption coefficient.

The weak tail of the absorption onset originates from defects and impurities. In this range, the absorption coefficient is independent of the excitation energy. In the medium range of the absorption onset, the absorption coefficient is dependent on the structural disorder of the material and shows an exponential dependence on the excitation energy. This range is referred to as Urbach tail. In the high absorption range, the absorption coefficient is dependent on the optical band gap of the material.

In literature, there a several models relating the energy-dependent absorption coefficient to the optical band gap ΔE^{opt}. A summary on the different models reported in literature is given by López et al.[151] Tauc proposed the following equation for the evaluation of the band gap energy:

$$\alpha h\upsilon \approx (h\upsilon - \Delta E^{opt})^p \tag{4.8}$$

The Tauc model is based on the assumptions that all matrix elements for the transitions upon photon absorption are equal, ΔE^{opt} corresponds to the energy of an indirect ($p = 2$) or direct $\left(p = \frac{1}{2}\right)$ allowed transition between the valence and conduction band, and that the electron states in the bands near the band gap exhibit a parabolic distribution $N(E) \sim E^{1/2}$ as for crystalline semiconductors. The optical band gap ΔE^{opt} can be derived by plotting $\sqrt{\alpha h\upsilon}$ or $(\alpha h\upsilon)^2$ on the ordinate against the energy $h\upsilon$ on the abscissa and extrapolating the linear part of the absorption onset to the abscissa.

As the Tauc model was derived from the theory of crystalline solids, for which the description of the electronic structure by bands is applicable, it is questionable whether this concept can be extended to amorphous and microcrystalline samples or metal organic molecules. Due to the lack of lattice periodicity, the wave vector and the differentiation between direct and indirect transitions are not well-defined concepts for non-crystalline systems. Furthermore, wavelength-dependent scattering and reflectance can affect the amount of transmitted or reflected light and consequently the measured absorption.[152] Nevertheless, the Tauc model has been applied to determine the HOMO-LUMO separation of metal organic and organic molecular materials, e.g., tris(8-hydroxyquinolinato)aluminium in solution,[153] and as thin film sample,[154] thin film samples of Cu, Ni, and Zn complexes,[155] polyallyl diglycol carbonate polymer films,[156] as well as triphenylene and perylene derivatives[157].

The reason for the widespread use of the Tauc model for the determination of the optical band gap ΔE^{opt} from optical absorption and diffuse transmittance spectra is probably the low technical demand for these measurements and the simplicity of the data analysis. When applying the Tauc model on of metal organic and organic molecular materials, it has to be considered that there is so far no theoretical proof for the applicability of this theory. Furthermore, the binding energy of the valence exciton present upon a HOMO-LUMO transition in a molecular material or valence band maximum to conduction band minimum transition in an inorganic semiconductor is not taken into consideration by this approach.

4.4.1.2 Cyclic voltammetry

Cyclic voltammetry is measured in a three-electrode setup with a reference electrode, working electrode, and counter electrode. To ensure sufficient conductivity, an electrolyte is added to the solution of the analyte. As the solution is not stirred during the experiment, the electrochemical reactions are diffusion-controlled. Common materials for the working electrode are noble metals like gold and platinum or glassy carbon. Counter electrodes may consist of the same materials as the working electrode. The reference electrode that is an electrode of the second type, e.g., a saturated calomel electrode, maintains a constant potential.[158]

The electrochemical reactions, take place at the working electrode, as the potential between the working electrode and the reference electrode is varied linearly with time. An electrochemical oxidation/reduction corresponds to the abstraction of an electron from the HOMO/addition of an electron to the LUMO of the analyte. The current is measured between the working electrode and the counter electrode. In a measurement cycle at first, no potential is applied between the working electrode and the reference electrode. As the potential is increased linearly with time during the forward scan, reduction or oxidation will occur at a certain point. A current will flow between the working electrode and the counter electrode in case there are reducible or oxidizable analytes in the system. After the concentration of reducible or oxidizable analyte is depleted, the current will decrease. As the potential is decreased in the reverse scan, the reduced or oxidized analyte will be re-oxidized or re-reduced, which yields a current of reverse polarity to before. The obtained data is plotted in a current-voltage diagram and provides information about redox potentials and electrochemical reaction rates. The measurement setup as well as idealized voltage-time and current-potential diagrams are depicted in Figure 47. Further details on cyclic voltammetry and mechanistic considerations are given in literature [158].

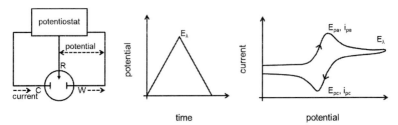

Figure 47: Left: Measurement setup for a cyclic voltammetry experiment.[158] C = counter electrode, R = reference electrode, W = working electrode. Middle: Applied potential vs. time. Right: Typical cyclic voltammetry diagram. E_{pc} = cathodic spike potential, i_{pc} = cathodic spike current, E_{pa} = anodic spike potential, i_{pa} = anodic spike current, E_{λ} = reversal potential.

Relative electrode potentials can be assessed by cyclic voltammetry. According to IUPAC convention, the relative electrode potentials are referenced to the standard hydrogen electrode as zero point. The electron affinity (EA) and the ionization potential (IP) can be derived from the relative electrode potentials according to an empirical correlation suggested in literature.[159, 160] For aqueous systems, the electrode potential of the saturated calomel electrode (SCE) is converted into electronic energies by adding a constant, which is in the range of 4.3–4.8 eV, with 4.4 eV being used most frequently.[159] Therefore, a scale of 4.4 eV SCE energy level below vacuum was used herein.

$$EA^{CV}(eV) = E_{onset}^{red}(vs.\ SCE) + 4.4 \qquad (4.9)$$

$$IP^{CV}(eV) = E_{onset}^{ox}(vs.\ SCE) + 4.4 \qquad (4.10)$$

Onset potentials versus the potential of ferrocene as internal standard are converted to the SCE scale.[159, 160]

$$E\ (SCE) = E\ (FeCp_2^+/FeCp_2) + 0.1588\ eV \qquad (4.11)$$

Oxidation and reduction potentials of different copper species have been studied in literature previously.[161-163] For CuCl and CuBr complexes with different N-aromatic and amine ligands in acetonitrile, oxidation potentials in the range of E_{pa} (SCE) = 0.32–(−0.35) V were found.[163] A study of Cu(I) complexes ($[Cu(PPh_3)_n(MeCN)_{4-n}]^+$ with n = 0–4) in dimethylformamide revealed that the oxidation potential for the oxidation of Cu(I) to Cu(II) ($[Cu(PPh_3)_n(MeCN)_{4-n}]^+ \rightarrow [Cu(PPh_3)_n(MeCN)_{4-n}]^{2+} + e^-$) increases with an increasing number of triphenylphosphine ligands in the range of E_{pa} (SCE) = 0.60–0.92 V for n = 0–4. The electrochemical reduction of the complexes in the series was also investigated. The shape of the cathodic wave indicated the irreversible deposition of copper metal on the electrode for all of the complexes.[162]

Advantageously, the experimental setup for cyclic voltammetry measurements is of low technical demand except the measures taken to run the experiment under inert gas conditions. Drawbacks for the determination of the IPs and EAs of Cu(I) complexes by cyclic voltammetry is the potential occurrence of dissociation and rearrangement reactions in solution as described previously and that the reduction potential cannot be assessed in case of irreversible reduction of Cu(I) to Cu(0) as described for Cu(I) complexes in literature. Furthermore, the often profound differences between photophysical properties of Cu(I) in solution and solid phase raise the question whether data obtained from dissolved samples is representative for amorphous thin films.

4.4.1.3 Photoelectron spectroscopy in air (PESA)[††]

An alternative method for the assessment of the ionization potential of metals, semiconductors, and organic materials like luminescent Cu(I) complexes, is photoelectron spectroscopy in air (PESA). The measurement principle is based on an open counter[164] that consists of a cylindrical cathode and a loop-shaped anode, between which a voltage is applied. There are two grids that suppress the detection of positive ions and successive or continuous gas discharge. Low-energy electrons emitted from a sample upon UV irradiation with a xenon lamp form O_2^- ions with the oxygen in the air. The O_2^- ions are accelerated towards the anode and detected. The measurement setup for photoelectron spectroscopy in air is depicted in Figure 48. The square root or cube root of the count rate is commonly plotted against the photon energy of the incident UV light. The onset of this curve is interpreted to be related to the ionization potential of the sample.

[††] The PESA measurements reported herein were conducted by Dipl.-Chem. Tanja Pürckhauer at the Light Technology Institute (LTI) at KIT.

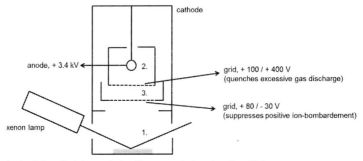

1. Irradiation with photons, emission of low energy electrons, formation of O_2^-
2. Gas discharge
3. Quenching and suppressing of excessive discharge (square pulse applied to grids)

Figure 48: Schematic depiction of the measurement setup for photoelectron spectroscopy in air.

Open questions concerning the PESA measurements remain regarding a possible oxidation of the sample surface, incorporation of impurities such as surface-bound oxygen and water, an incorrect vacuum level, the impact of the Xe lamp, as well as the significance of the derived values for the ionization potential of a sample that has not been exposed to air like a thin film in an OLED.

4.4.1.4 Photoelectron spectroscopy (PES)

X-ray photoelectron spectroscopy (XPS) and ultraviolet photoelectron spectroscopy (UPS) can be used to study the occupied electronic levels of a sample in analogy to x-ray emission spectroscopy (4.3.1). In contrast to RIXS and optical spectroscopy, no core or valence exciton binding energies have to be taken into account. The principle of photoelectron spectroscopy is based on the photoelectric effect, which was first described by Heinrich Hertz and Wilhelm Hallwachs in 1887.[165] In an XPS or UPS measurement, the sample is irradiated by photons of a defined energy. Possible photon sources are x-ray sources generating Mg K_α or Al K_α radiation for XPS, helium gas discharge lamps generating He I and He II radiation for UPS, or a synchrotron radiation facility, in which case, the abbreviation PES is often used.

For XPS, the spectrum is calibrated in energy by reference core level and Auger energies from reference measurements of, e.g., gold, silver, and copper, while for UPS, reference measurements of the Fermi edge of a metal sample are used. The energy of the electrons of this metal sample at the Fermi edge is set to $E_{kin} = h\nu$ and the spectrum of the sample is shifted accordingly. The energy scale is now referenced to the Fermi energy of the sample. Emitted photoelectrons of the sample have binding energies according to $E_{bin} = h\nu - E_{kin}$ and the secondary electron cutoff for electrons ejected from the sample appears at $E_{bin} = h\nu - \Phi_s$. By determining the binding energy of the cutoff the work function of the sample can be assessed.

The probability for the emission of a photoelectron can be derived from Fermi`s Golden Rule as for x-ray absorption and emission (see 4.2.1.2 and 4.3.1): Some of the photons are absorbed by the atoms in the sample and the ejected electrons are detected as a function of their energy. The inelastic mean free path of electrons in a solid is short – in the range of 2–30 Å, depending on their kinetic energy and the matrix – therefore, photoelectron spectroscopy is a very surface-sensitive method. The relationship between the inelastic mean free path and the kinetic energy of electrons in a solid is approximated by the "universal curve" shown in Figure 49.[166] There are also more modern, theoretical approaches for the calculation of the inelastic mean free paths of electrons in a solid. Tougaard et al. calculated the inelastic mean free paths based on the Monte Carlo model[167] and Penn et al. used a modified form of the Bethe equation[168],[169, 170] for example. Upon excitation, the system is excited from its initial state $|\Psi_i\rangle$ into the final state $\langle\Psi_f|$. If the wave function of the initial and final state is approximated by products of the single-electron wave functions of all N electrons in the system and if the perturbation is assumed to only interact with the wave function of the excited electron, the probability for the emission of a photoelectron can be described based on equation (4.1).[109]

$$
\begin{aligned}
\omega_{i \to f} &\propto \left|\langle\Phi_f|\hat{H}_s|\Phi_i\rangle\right|^2 \left|\langle\Phi_f(N-1)|\Phi_i(N-1)\rangle\right|^2 \times \\
&\quad \delta\left[E_{kin} + E_f(N-1) - E_i(N) - h\nu\right]
\end{aligned}
\tag{4.12}
$$

Φ_i and Φ_f represent the initial and final state of the excited electron and $\Phi_i(N-1)$ and $\Phi_f(N-1)$ the initial and final state of the remaining electrons, respectively. $\hat{H}_s$ represents the perturbation of the one-electron system by the electromagnetic field of the photon. E_{kin} is the kinetic energy of the ejected electron. $E_i(N)$ and $E_f(N-1)$ are the initial energy of the unperturbed N electron system and the final energy of the remaining $(N-1)$ electron system.

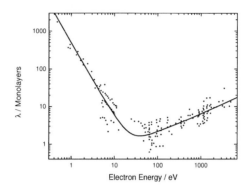

Figure 49: "Universal curve" for the inelastic mean free path as a function of the kinetic energy of electrons in a solid.[166]

The δ-function in equation (4.12) ensures the energy conservation during the photoemission process. Applying Koopmans` theorem (frozen orbital approximation) $\left|\Phi_f(N-1)\right\rangle$ and $\left|\Phi_i(N-1)\right\rangle$ can be assumed to be equal. By substituting $E_i(N) = E_i(N-1) + E_{i,j}$ and assuming $E_i(N-1) = E_f(N-1)$, the equation for the energy conservation can be written as follows:

$$E_{kin} \approx h\nu + E_{i,j} = h\nu - E_{bin,K} \tag{4.13}$$

$E_{bin,K}$ is referred to as Koopmans` binding energy, which equals the negative orbital energy of the ejected photoelectron $E_{i,j}$.

In general, Koopmans` binding energies $E_{bin,K}$ are not in agreement with the measured binding energies E_{bin}, for several reasons.

First, $\left|\Phi_f(N-1)\right\rangle \neq \left|\Phi_i(N-1)\right\rangle$, which means that the remaining electronic system, which contains a hole in the final state, relaxes. This "screening" process causes E_{bin} to be smaller than $E_{bin,K}$. Second, there are further effects influencing the binding energy E_{bin} in bound atoms with respect to unperturbed free atoms, which can be classified into initial state and final state effects.[171] A schematic presentation of initial and final state effects is given in Figure 50. In particular, initial state effects are intra-atomic spin orbit splitting of orbitals with angular momentum quantum number larger zero $(l > 0)$ and Coulombic interactions. Coulombic interactions refer to the influence of the charge density within the probed atom on the binding energy E_{bin}. The charge density is controlled by the electronic structre of the probed atom (interatomic) and the local chemical environment around the probed atom (intra-atomic). Based on these effects, it is possible to conclude which elements are present in a sample from the measured photoelectron energies and to obtain information on, e.g., the oxidation state and binding atoms. Final state screening processes are attributed to excited-state polarization or rearrangement effects. Core hole-induced polarization is due to the charge and spin introduced in the final state upon emission of the photoelectron. This also can lead to multiplet splitting in case of unpaired valence electrons in the probed atom. Furthermore, there are rearrangement effects like shake-up satellites (excitation of valence electrons to unoccupied orbitals) and shake-off satellites (excitation of valence electrons to vacuum) and plasmon generation. As the emission of Auger electrons along with fluorescence are the primary decay channels upon core hole generation, Auger lines can be observed together with photoelectron emission for all elements containing at least three electrons.[171]

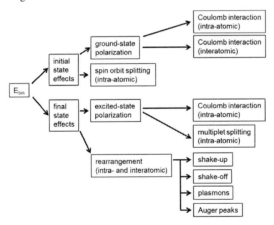

Figure 50: Initial and final state effects influencing the binding energy E_{bin} in bound atoms with respect to unperturbed free atoms.[171]

4.4.1.5 Inverse photoemission spectroscopy

Whereas PES can be used to study the occupied electronic levels, inverse photoemission spectroscopy (IPES) enables the study of unoccupied electronic levels, in analogy to XAS.

As for UPS, no core or valence exciton binding energies have to be taken into account for IPES. IPES can be measured in a wide range of photon energies. For the herein present IPES measurements, electrons with low kinetic energies (in the range of 5–20 eV) emitted from an electron gun were used. These electrons are injected into the sample and occupy previously unoccupied high lying electronic states. These excited states may decay via radiative transitions to other, lower lying unoccupied electronic states. As for direct photoemission, this processes can be described by Fermi`s Golden Rule (see equations (4.1) and (4.2)). Note, however, that IPES involves $(N + 1)$ electrons, and thus none of the initial and final states are the same. For example, the final state $\langle \Psi_f |$ for IPES differs from that for PES. For IPES, there is an additional electron occupying a previously unoccupied electronic state and in PES there is a hole in a previously occupied electronic state. A spectrometer or a Geiger-Müller counter can be used as detector for IPES to detect photons in the low energy range.[172, 173] The cross section for IPES is much lower in comparison to PES due to the different phase space available for photon creation or electron creation. Assuming that the squares of the matrix elements for IPES and PES can be equated to each other, the ratio of the two cross sections is given by $r_{IPES/PES} = (\lambda_e / \lambda_p)$, where λ_e and λ_p are the electron an photon wavelengths.[174] At a typical energy of 10 eV, the ratio of the two cross sections is $r_{IPES/PES} = 10^{-5}$. Therefore, a high electron flux and a good detection efficiency to obtain IPES spectra with a good signal-to-noise ratio are required.

When combined, UPS and IEPS measurements can be used to determine the HOMO and LUMO energies with reference to the Fermi energy as well as the work function of a sample, and thus give the most complete picture of the electronic (surface) structure.

4.4.2 Results: Assessment of the HOMO and LUMO energies by optical spectroscopy, cyclic voltammetry, PESA, UPS, and IPES

4.4.2.1 Optical spectroscopy

The NHetPHOS **1**, **5**, and **6** were selected based on their different emission color ranging from blue (NHetPHOS complex **5**) over green (NHetPHOS complex **1**) to red (NHetPHOS complex **6**), to evaluate different methods for the assessment of the HOMO and LUMO energies of these materials. Molecular structures of NHetPHOS complexes **1**, **5**, and **6** are depicted in Figure 51.

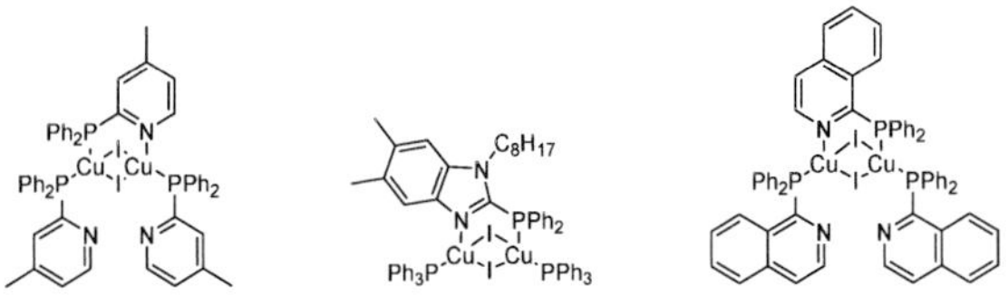

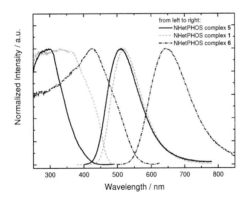

Figure 51: Molecular structures of NHetPHOS complexes **1**, **5**, and **6**.

Optical excitation and emission spectra for solid samples of NHetPHOS complexes **1**, **5**, and **6** are depicted in Figure 52. The features in the excitation and emission spectra are broad and unstructured, as for most Cu(I) complexes. The emission maxima of the NHetPHOS complexes **1**, **5**, and **6** are found at approximately 522 nm, 510 nm, and 645 nm, respectively. The low-energy onset of the excitation spectra shifts in correspondence to the maximum of the emission spectra, with NHetPHOS complex **5/6** exhibiting an excitation onset at higher/lower energies than NHetPHOS complex **1**.

Figure 52: Excitation and emission spectra of crystalline/film/powder samples of NHetPHOS complexes **1/5/6**.

In correspondence to equation (4.8), the optical band gap of the complexes was assessed by Tauc plots.

$$Ihv \approx \left(hv - \Delta E^{opt}_{HOMO-LUMO} \right)^2 \qquad (4.14)$$

The square root of the product of the normalized intensity of the excitation spectra depicted in Figure 52 and the respective excitation energy in eV $\sqrt{Ih\upsilon}$ was plotted against the excitation energy $h\upsilon$ in eV. The linear section of the graph was extrapolated to the baseline (see Figure 53). The derived HOMO-LUMO separations are given in Table 9. By drawing the steepest and flattest tangents to the baseline and the Tauc function, still consistent with the data, error bars for the determination of the HOMO-LUMO separation were estimated to be $\Delta E = \pm 0.05$ eV.

Table 9: Peak emission energies and HOMO-LUMO separations derived from optical excitation spectra according to Tauc.[175]

	NHetPHOS complex **1**	NHetPHOS complex **5**	NHetPHOS complex **6**
peak emission / nm	522 ± 5	510 ± 5	645 ± 5
peak emission / eV	2.4 ± 0.05	2.4 ± 0.05	1.9 ± 0.03
$\Delta E^{opt}_{HOMO-LUMO}$ / eV	2.6 ± 0.05	2.8 ± 0.05	2.1 ± 0.05

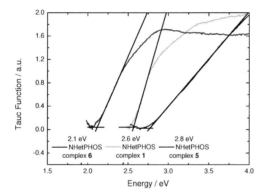

Figure 53: Estimation of the optical band gap of NHetPHOS complexes **1**, **5**, and **6** according to Tauc.[175] $\sqrt{Ih\upsilon}$ was plotted against energy in eV for the excitation spectra of crystalline/film/powder samples of NHetPHOS complexes **1/5/6**.

The relative order of the derived HOMO-LUMO separations, with NHetPHOS complex **5/6** exhibiting a bigger/smaller HOMO-LUMO separation than NHetPHOS complex **1**, correlates with the respective energy (color) of the photons emitted by the materials. A summarizing discussion of the obtained values for the HOMO-LUMO separations is given in chapter 4.4.2.5.

4.4.2.2 Cyclic voltammetry

Cyclic voltammetry measurements on Cu(I) NHetPHOS complexes **1**, **5**, and **6** were conducted in dichloromethane and are shown in Figure 54.

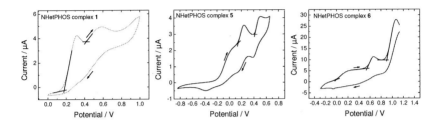

Figure 54: Cyclic voltammetry spectra of NHetPHOS complexes **1**, **5**, and **6** in dichloromethane with tetrabutylammonium hexafluorophosphate as electrolyte. The concentration of the analyte was approximately 10^{-3} M in dichloromethane and the scan rate was set to 0.05 V/s. The potential was measured with reference to ferrocene as internal standard. Oxidation onset potentials were determined from intercept of the tangents between the respective baselines and the signal current.

The scanned voltage was limited to a range in which the analyte was oxidized to avoid the irreversible reduction of Cu(I) to Cu(0). The concentration of the analyte was approximately 10^{-3} M in dichloromethane and the scan rate was 0.05 V/s.

The obtained cyclic voltammetry spectra differ from an idealized spectrum (see Figure 47), because the electrochemical reactions are not fully reversible for the Cu(I) NHetPHOS complexes. The cathodic spike current for the electrochemical oxidation reactions did exceed the anodic spike current for the re-reduction, i.e., $i_{pc} \neq i_{pa}$. Furthermore, there a two or three oxidation waves distinguishable for all three NHetPHOS complexes. A similar behavior was observed for other Cu(I) complexes, in which case the first oxidation wave was assigned to the Cu(I) complexes and the second wave to free ligands.

Table 10: Oxidation onset potentials versus the potential of ferrocene as internal standard and derived ionization potentials IP^{CV} relative to the vacuum level. By drawing the steepest and flattest tangents to the baseline and the signal current, still consistent with the data, errors bars for the determination of the onset potentials and the ionization potentials were estimated to be $\Delta E = \pm 0.05$ eV.

material	E_{onset}^{ox} / V	IP^{CV} / eV
NHetPHOS complex 1	**0.18**, 0.42 ($\pm$ 0.05)	4.7 $\pm$ 0.05
NHetPHOS complex 5	**0.15**, 0.42 ($\pm$ 0.05)	4.7 $\pm$ 0.05
NHetPHOS complex 6	**0.00**, 0.55, 0.90 ($\pm$ 0.05)	4.6 $\pm$ 0.05

Accordingly, the onset of the first oxidation wave was determined from the intercept of the tangents between the baseline and the signal current in the anodic scan. Sometimes, oxidation potentials are determined from the maximum of the first oxidation wave. In this case the oxidation potentials would approx. 0.10–0.20 V higher for the Cu(I) NHetPHOS complexes. By drawing the steepest and flattest tangents to the baseline and the signal current, errors bars for the determination of the onset potentials and the ionization potentials were estimated to be $\Delta E = \pm 0.05$ eV. A summarizing discussion of the IPs obtained by cyclic voltammetry is given in chapter 4.4.2.5.

4.4.2.3 Photoelectron spectroscopy in air[‡‡]

In cooperation with the Light Technology Institute (LTI) at KIT, PESA measurements on selected luminescent Cu(I) complexes were conducted by Dipl.-Chem. Tanja Pürckhauer. The measurements were performed on a RIKEN AC-2 system under atmospheric conditions. Film samples of NHetPHOS complexes **1**, **5**, and **6** were prepared on glass slides by drop-casting from toluene. The films were tempered on a hot plate at 60°C for 20 minutes. The data is shown in Figure 55 and the evaluation was conducted according to an empirical power law for the spectral yield Y near the threshold energy E^* in dependency of the incident energy E.[164, 176, 177]

$$Y \propto (E + E^*)^n, with\ n = 1, \frac{3}{2}, 2, and\ \frac{5}{2} \tag{4.15}$$

For metals, typically the square root version is used, while for classical inorganic semiconductors and molecular systems, the cube root version appears to be more appropriate.[164, 176, 177] To determine the ionization potential of a sample, the square root or cube root of the signal is plotted versus the excitation energy and the onset is determined from the intercepts of tangents fitted to the baseline and the signal.

Thus, for NHetPHOS complexes **1**, **5**, and **6**, a cube-root representation of the PESA signal was used to determine the ionization potential. Based on the variance of the derived values for different film samples of the same material, an error of ±0.2 eV is estimated for the

---

[‡‡] PESA measurement on chosen luminescent Cu(I) complexes were conducted by Dipl.-Chem. Tanja Pürckhauer in cooperation with the Light Technology Institute (LTI) at KIT,

ionization potentials derived by PESA measurements. For complex **1**, **5**, and **6**, values of IP^{PESA} = 5.3, 5.3, and 5.2 (±0.2) eV were derived, respectively. A summarizing discussion of the IPs obtained by PESA is given in chapter 4.4.2.5.

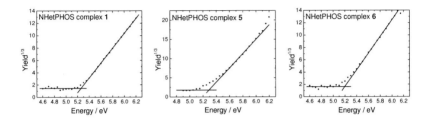

Figure 55: Determination of the ionization potentials IP^{PESA} of NHetPHOS complexes **1**, **5**, and **6** using photoelectron spectroscopy in air (PESA) and a cube-root representation of the PESA signal. PESA measurements were conducted on film samples on glass substrates.

4.4.2.4 Photoelectron and inverse photoemission spectroscopy

To study NHetPHOS complex **1**, **5**, and **6** with XPS, UPS, and IPES, films of each material were deposited on gold-coated silicon wafers via spin-coating from solution. As charging under exposure to x-ray or ultraviolet radiation and electrons was suspected to be problematic due to the low conductivity of the Cu(I) complexes, films with different thicknesses for each material were prepared. The influence of charging of the films on the spectra can be assessed, as charging can be expected to be bigger for thicker films. Subsequently, XPS, UPS and IPES were measured on all samples. For clarity, all XPS photoemission and Auger lines of the elements present in the spin-coated films of the investigated Cu(I) complexes are depicted in Figure 56.

XPS survey spectra of the gold-coated silicon wafer substrate and three films with different thicknesses of NHetPHOS complex **1**, spin-coated from toluene, are shown in Figure 57. The inset in Figure 57 shows detail spectra of the Au $4f_{5/2, 7/2}$ peaks.

Detail spectra for each element present in the spin-coated films of the investigated Cu(I) complexes were measured to investigate the elemental composition of the different films (Figure 57).

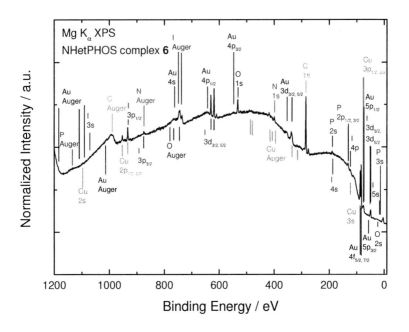

Figure 56: Mg K$_\alpha$ XPS spectrum of a Cu(I) NHetPHOS complex **6**, with markers indicating the position of all photoemission and Auger lines for the elements present in the sample.

The spectra of the gold-coated silicon wafers substrates reveal the presence of oxygen and carbon at the gold surface of the substrate. For the Cu(I) complexes, there are also signals that stem from oxygen, in addition to the signals from the elements of which the Cu(I) complexes consist (copper, iodine, nitrogen, carbon, and phosphorus).

The shape and position of the Cu $2p_{1/2}$, I $3d_{3/2, 5/2}$, N 1s, C 1s, and the P $2p_{1/2, 3/2}$ peaks differ for the different films of NHetPHOS complexes **1**, **5**, and **6**. Detail spectra of NHetPHOS complex **1** are exemplarily shown in Figure 58. A shift to lower binding energies of up to 0.4 eV for the Cu $2p_{1/2}$, I $3d_{3/2, 5/2}$, and C 1s signal could be attributed to a decrease in charging of the films with decreasing films thickness. Another possible explanation for the shift of the Cu $2p_{1/2}$, I $3d_{3/2, 5/2}$, and C 1s signal is a thickness-dependent change of the energy level alignment at the interface and/or formation of an interface dipole, causing a shift in XPS and UPS, as has previously been reported for Cu(II) phthalocyanine on polycrystalline and crystalline Au[178] and 3,4,9,10 perylenetetracarboxylic dianhydride (PTCDA) on ITO[179].

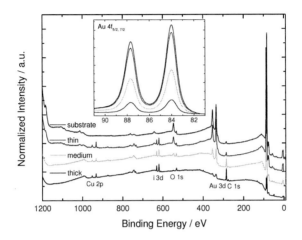

Figure 57: Mg Kα XPS survey spectra of the gold-coated silicon wafer (substrate), and films with different thickness (thick, medium, thin) of NHetPHOS complex **1** spin-coated on the substrate. The inset shows detail spectra of the Au 4f5/2, 7/2 peaks.

It is not possible to distinguish between charging upon exposure to photons/electrons and an inherent surface charge, based on the herein presented data. However, the same shift cannot be observed for the N 1s signal, and the peak maximum of the P 2p $_{1/2, 3/2}$ signal is shifted to higher binding energies. This suggests that effects like charging, energy level alignment at the interface, and formation of an interface dipole alone cannot explain the differences in peak shape and position of the thick, medium, and thin films. Additional effects might be decomposition of the Cu(I) complexes in due to vacuum and/or x-ray exposure, and binding of oxygen or water impurities in the glove box to the sample surface.

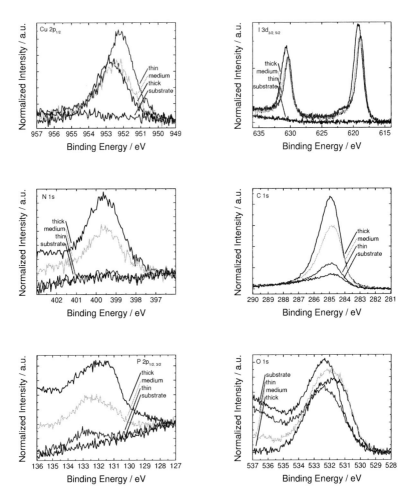

Figure 58: Mg K_α detail spectra of the gold-coated silicon wafer (substrate) and spin-coated films with different thicknesses (thick, medium, thin) of NHetPHOS complex **1** on the substrate as well as the C 1s and O 1s detail spectra of the substrate.

Subsequent to the characterization of the film of NHetPHOS complexes **1**, **5**, and **6** by XPS, UPS measurements were conducted on all film samples and the Au substrate to study the occupied molecular orbitals and the work functions of the surfaces. Detail spectra of the secondary electron cutoff and the highest occupied orbitals of the films are shown in

Figure 59. Furthermore, the secondary electron cutoff and the Fermi edge of the gold substrate are depicted in Figure 59.

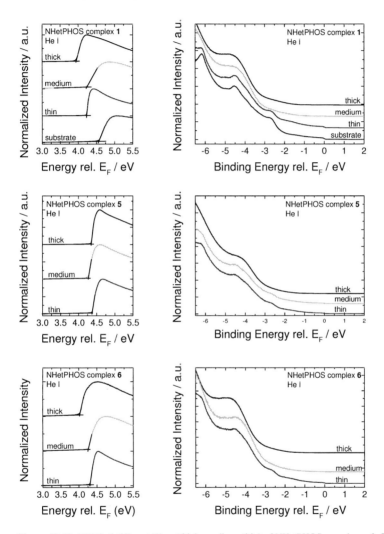

Figure 59: He I UPS of different films (thick, medium, thin) of NHetPHOS complexes **1**, **5**, and **6**. The detail spectra show the secondary electron cutoff (left) and high lying occupied orbital+s (right).

The secondary electron cutoff can be used to determine the work function of the samples from the intercept of the tangents between the baseline and the onset. The derived values for the work function Φ_s of the different samples (thick, medium, thin) of NHetPHOS complex **1**, **5**, and **6** are listed in Table 11. For the substrate a work function of $\Phi_s = 4.6$ eV was derived. For polycrystalline and crystalline gold surfaces, work functions in the range of $\Phi_s = 5.1$–5.47 eV[180, 181] were derived. The work function of the herein used gold substrate is significantly lower. This might be due to the oxygen and carbon present at the gold surface of the gold substrate. It has been shown that monolayers of organic material on Au(111) can exhibit a work function as low as $\Phi_s = 4.30$ eV, which is 0.9–1.0 eV lower than typical values reported for Au(111) surfaces.[182] The work function of graphite surfaces has been reported to be in the range of $\Phi_s = 4.6$–4.7 eV.[183-185]

By drawing the steepest and flattest tangents to the baseline and the onset, still consistent with the data, errors bars for the determination of the work function were estimated to be $\Delta E = \pm 0.05$ eV. The work functions for the thin and medium films of NHetPHOS complexes **1**, **5**, and **6** are in the range of 4.2–4.4 eV and are in agreement with each other within 0.10 eV, which is within the error bar, whereas there is a significant difference between the thick and medium/thin films of up to 0.3 ± 0.10 eV.

Table 11: Work function Φ_s for all samples (thick, medium, thin) of NHetPHOS complexes **1**, **5**, and **6**. A work function of $\Phi_s = 4.6 \pm 0.05$ eV was derived for the surface of the substrate.

Φ_s / eV	thick	medium	thin
NHetPHOS complex **1**	4.0 ± 0.05	4.2 ± 0.05	4.2 ± 0.05
NHetPHOS complex **5**	4.4 ± 0.05	4.3 ± 0.05	4.4 ± 0.05
NHetPHOS complex **6**	4.0 ± 0.05	4.3 ± 0.05	4.3 ± 0.05

By comparison of the UPS detail spectra of the upper occupied orbitals for the different films (thick, medium, thin) of NHetPHOS complexes **1**, **5**, and **6**, distinct differences can be identified. For the spectrum of the gold substrate (Figure 59, upper graph), the metal Fermi cutoff is clearly visible. Furthermore, the spectrum of the gold substrate shows three distinct peaks at approx. 2.7, 4.5, and 6.2 eV binding energy w.r.t. the Fermi level. For the thin films of NHetPHOS complexes **1**, **5**, and **6** (Figure 59, upper, middle and lower graph), the metal Fermi cutoff is reduced in intensity but still visible. The intensity of the metal Fermi cutoff is even more reduced for the medium films and is no longer visible for the thick films. Hence, the three peaks, observed in the spectrum of the gold substrate, are also visible in the spectra of the thin films and partially visible in the spectra of the medium films. These peaks cannot be observed in the spectra of the thick films.

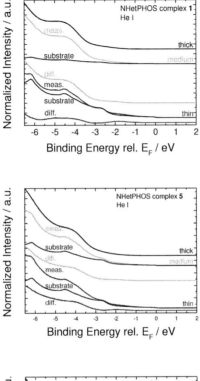

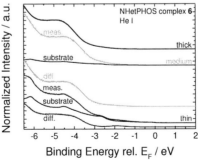

Figure 60: He I UPS of different films (thick, medium, thin) of NHetPHOS complexes **1**, **5**, and **6**. The detail spectra show the high lying occupied orbitals: Measured spectra and difference spectra difference spectra (diff.) [= measured spectra (meas.) – substrate] are shown.

The spectra of NHetPHOS complexes **1**, **5**, and **6** show a signal onset in the region between 0.5–2.0 eV binding energy w.r.t. the Fermi level, which can probably be attributed to the HOMOs of the complexes.

There is a shift to higher binding energies in the signal onset up to 0.3 eV for the thick films in comparison to the medium and thin films. For even thicker films, a shift to higher binding energies of several electron volts was observed. To further evaluate the differences in the signal onset for the different film spectra of each NHetPHOS complex, difference spectra were obtained by normalizing the signal of the Au substrate, such that the signal intensity of the Fermi edge in the Au substrate spectrum matches the signal intensity of the Fermi edge in the film spectra, and subtracting the normalized Au spectra from the film spectra. Measured (meas.) and difference spectra (diff.) are depicted in Figure 60.

This approach has shortcomings, as there are effects that might emerge in the spectra: The inelastic mean free path of electrons in solids is dependent on the kinetic energy of the electrons and there might be interactions between the substrate and functionalities of the Cu(I) complexes. For example, the binding of pyridine on gold has been studied extensively and might occur at the interface of the gold and the metal organic film.[186-188] Therefore, the measured spectrum of a film on a substrate is not necessarily a linear combination of the spectra of the pristine material in the film and the substrate.

The difference in the work function for some of the thick films and the onset of the signal count might be due to charging of the films under exposure to ultraviolet light and the subsequent emission of electrons from the sample. As the Cu(I) complexes exhibit a low conductivity, a partial positive charge might be created at the sample surface, which would lead to a shift of the spectrum to lower kinetic energies and the observation of a lower work function.

The overall good agreement in the work function of the medium and thin films and the more distinct difference for the work function of the thick films could indicate that charging is not as pronounced for the medium and the thin films as for the thick films, although charging cannot be completely excluded even for the medium and thin films. Another possible explanation for the differences in the work function is a thickness-dependent variation of the energy level alignment at the interface[178] and/or formation of an interface dipole.[178, 179]

As the onset in UPS is shifted by several electron volts for thicker films, a thickness-dependent variation of energy level alignment at the interface and/or formation of an interface dipole might not fully explain the observed shift and other factors like charging should be taken into account.

Following the characterization of the different films of NHetPHOS complexes **1**, **5**, and **6** by XPS and UPS, IPES measurements were conducted on all samples to study the unoccupied orbitals. Furthermore, the conduction band of the Au substrate was characterized (see Figure 61).

For the spectrum of the gold substrate (Figure 61, upper graph), the metal Fermi cutoff is clearly visible. The spectrum of the gold substrate shows a feature with constant intensity in the range of 1.0–3.4 eV w.r.t. the Fermi level and an approximately linear increase in intensity in the range of 3.4–6.4 eV w.r.t. the Fermi level. For the thin films of NHetPHOS complexes **1**, **5**, and **6** (Figure 61, upper, middle and lower graph), the metal Fermi cutoff is reduced in intensity but still visible. The intensity of the metal Fermi cutoff is even more reduced for the medium films and is no longer visible for the thick films. Hence, the feature with constant intensity in the range of 1.0–3.4 eV w.r.t. the Fermi level, observed in the spectrum of the gold substrate, is also visible in the spectra of the thin and medium but not of the thick films. The spectra of NHetPHOS complexes **1**, **5**, and **6** show a signal onset in the range between 1.5–2.8 eV w.r.t. the Fermi level, which can probably be attributed to the LUMOs of the complexes.

To evaluate the differences in the signal onset for the different film spectra of each NHetPHOS complex, difference spectra were obtained by normalizing the signal of the Au substrate, such that the signal intensity of the Fermi edge in the Au substrate spectrum matches the signal intensity of the Fermi edge in the film spectra, and subtracting the normalized Au spectra from the film spectra. Measured (meas.) and difference spectra (diff.) are depicted in Figure 62.

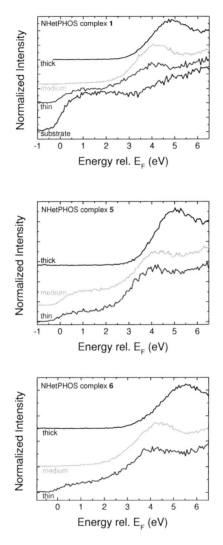

Figure 61: IPES spectra of an gold substrate and different films (thick, medium, thin) of NHetPHOS complexes **1**, **5**, and **6**.

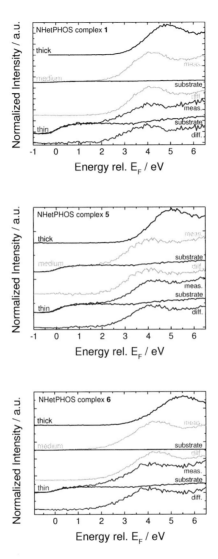

Figure 62: IPES spectra of an gold substrate and different films (thick, medium, thin) of NHetPHOS complexes **1**, **5**, and **6**. Measured spectra (meas.) and difference spectra (diff.) [= measured spectra – substrate] are shown.

The measured IPES spectra have a lower signal-to-noise ratio than the UPS spectra as the number of IPES scans on each sample is limited by radiation damage and IPES is an inherently low-count-rate technique. The signal onset for the thick films of the Cu(I) complexes is shifted to higher energies by approximately 1 eV, which suggests that charging of the thick films occurs upon exposure to electrons with low kinetic energies. Therefore only, the difference spectra of the medium and thin films of NHetPHOS complex **1**, **5**, and **6** will be compared in the following.

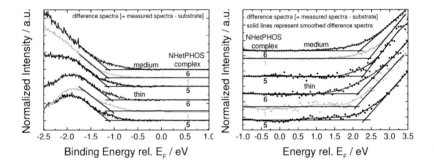

Figure 63: UPS (left) and IPES (right) detail spectra of the signal onset.

As for UPS, the approach of subtracting the signal from Au substrate has shortcomings as the inelastic mean free path of electrons in solids is dependent on the kinetic energy of the electrons and the measured spectrum of a film on a substrate is not necessarily a linear combination of the spectra of the pristine material in the film and the substrate. However, the comparison of the UPS and IPES difference spectra can be a suitable way of data analysis for the medium and the thin films of NHetPHOS complex **1**, **5**, and **6**, as the error introduced by the subtraction should be similar for all samples. The intercept of the tangents between the baseline and the signal onset is commonly used to determine the HOMO and LUMO energies of organic semiconductors by UPS and IPES.[189] For the UPS spectra, there is no significant difference in the energy position of the onset between the medium and thin films of NHetPHOS complex **1**, **5**, and **6**. The intercept of the baseline and the signal onset yields a HOMO binding energy of $E_{HOMO,F}^{UPS} = 1.1 \pm 0.05$ eV w.r.t. the Fermi level. Error bars for the determination of the HOMO binding energy were estimated ($\Delta E = \pm 0.05$ eV) by drawing the steepest and flattest tangents to the baseline and the signal onset, still consistent with the data. The IPs can be assessed by adding the work function of the respective sample derived from the secondary electron cutoff. This results in IPs^{UPS} in the range of 5.3–5.5 ± (0.2) eV. EAs were derived from the signal onset in the IPES difference spectra and the work function. As the signal-to-noise ratio for IPES spectra is much lower than for the UPS spectra, the IPES difference spectra were smoothed, and an error of ±0.1 eV was estimated for the EAs. The derived LUMO energies w.r.t. the Fermi level for the medium and thin films of NHetPHOS complex **1**, **5**, and **6** are in the range of $E_{LUMO,F}^{IPES} = 2.1$–$2.4 \pm (0.1)$ eV. This results in an EA^{IPES} of 1.8–2.2 ± (0.15) eV and HOMO-LUMO separation of $\Delta E_{HOMO-LUMO}^{UPS+IPES} = 3.2$–$3.5 \pm (0.15)$ eV for the medium and thin films of NHetPHOS complex **1**, **5**, and **6**. UPS and IPES results are summarized in Table 12.

Table 12: Work function Φ_s, IP, $E_{LUMO,F}^{PES}$ for the medium and thin film samples of NHetPHOS complex **1**, **5**, and **6**.

Φ_s	Φ_s / eV	IP^{UPS} / eV	$E_{LUMO,F}^{IPES}$	EA^{IPES} / eV	$\Delta E_{HOMO-LUMO}^{UPS+IPES}$ / eV
		NHetPHOS complex **1**			
medium	4.2 ± 0.05	5.3 ± 0.1	2.4 ± 0.1	1.8 ± 0.15	3.5 ± 0.15
thin	4.2 ± 0.05	5.3 ± 0.1	2.1 ± 0.1	2.1 ± 0.15	3.2 ± 0.15
		NHetPHOS complex **5**			
medium	4.3 ± 0.05	5.4 ± 0.0	2.3 ± 0.1	2.0 ± 0.15	3.4 ± 0.15
thin	4.4 ± 0.05	5.5 ± 0.1	2.3 ± 0.1	2.1 ± 0.15	3.4 ± 0.15
		NHetPHOS complex **6**			
medium	4.3 ± 0.05	5.4 ± 0.1	2.2 ± 0.1	2.1 ± 0.15	3.3 ± 0.15
thin	4.3 ± 0.05	5.4 ± 0.1	2.1 ± 0.1	2.2 ± 0.15	3.2 ± 0.15

Below, a summarizing discussion of the HOMO-LUMO separations, IPs, and EAs obtained by UPS and IPES can be found.

4.4.2.5 Summarizing discussion of the HOMO and LUMO energies by optical spectroscopy, cyclic voltammetry, PESA and electron spectroscopy

Upon comparison of the different values for the IPs, EAs, and HOMO-LUMO separations, it is evident that there are differences between the energies derived by optical spectroscopy, cyclic voltammetry, PESA, and electron spectroscopy. The results of the different techniques are summarized in Table 13 and depicted in Figure 64.

Table 13: Summary of the HOMO-LUMO separations and HOMO and LUMO energies w.r.t. the vacuum level derived by optical spectroscopy, cyclic voltammetry, PESA, UPS, and IPES.

NHetPHOS complex	1	5	6
$\Delta E^{opt}_{HOMO-LUMO}$ / eV	2.6 ± 0.05	2.8 ± 0.05	2.1 ± 0.05
IP^{cv} / eV	4.7 ± 0.05	4.7 ± 0.05	4.6 ± 0.05
IP^{PESA} / eV	5.3 ± 0.20	5.3 ± 0.20	5.2 ± 0.20
IP^{UPS} / eV *	5.3 ± 0.10	5.5 ± 0.10	5.4 ± 0.10
EA^{IPES} / eV	2.1 ± 0.15	2.1 ± 0.15	2.2 ± 0.15
$\Delta E^{UPS+IPES}_{HOMO-LUMO}$ / eV *	3.2 ± 0.15	3.4 ± 0.15	3.2 ± 0.15

* For IP^{UPS}, EA^{IPES}, and $\Delta E^{PES}_{HOMO-LUMO}$ only the values of the thin films of the Cu(I) complexes, for which charging can be assumed to be less pronounced, are presented.

Within each method, no significant differences for the IPs of NHetPHOS complexes **1**, **5**, and **6** are found. The values derived by PESA and UPS are in agreement with each other. However, cyclic voltammetry predicts up to 0.8 ± 0.25 eV higher-lying HOMOs than PESA and UPS. The reason for the difference in the values from cyclic voltammetry could be a systematical error in the conversion of electrochemical redox potentials to HOMO energies with regards to the different scales that are used for conversion in literature. With a SEC potential of 4.8 eV w.r.t. the vacuum level, the deviation would be reduced to 0.4 ± 0.25 eV. In case oxidation potentials were determined from the maximum of the first oxidation wave instead of the onset, the deviation would be further reduced to approx. 0.20–0.30 (± 0.25 eV). A further reason for the difference could be the formation of new chemical species by dissociation or rearrangement reactions in solution. Furthermore, charging of the thin films and the degradation of the Cu(I) complexes in UHV and binding of oxygen or water impurities in the glove box to the sample surface have to be taken into consideration. For PESA, a possible oxidation of the sample surface and incorporation of impurities such as surface-bound oxygen and water might impact the observed HOMO energy.

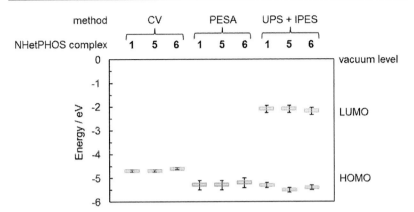

Figure 64: HOMO and LUMO energies w.r.t. the vacuum level derived by different methods. Solution samples of the NHetPHOS complexes were measured by cyclic voltammetry. PESA was measured on film samples in air and PES/IPES results were derived for thin film samples of the materials.

The HOMO-LUMO separations derived by Tauc`s method from optical excitation spectra of NHetPHOS complexes **1**, **5**, and **6** differ significantly from the values obtained by UPS and IPES for the thin films of the materials and are up to 1.3 ± 0.2 eV smaller. While there are significant differences of up to 0.7 ± 0.1 eV among the HOMO-LUMO separations derived for NHetPHOS complexes **1**, **5**, and **6** by optical spectroscopy, there is no significant difference for HOMO-LUMO separations of the thin films obtained by UPS and IPES. Judging from the photophysical characterization by Tauc plots and the emission spectra, NHetPHOS complex **5**/complex **6** would be expected to exhibit a higher/smaller HOMO-LUMO separation than NHetPHOS complex **1**.

The valence exciton binding energies have to be taken into account for the energy differences obtained by optical spectroscopy, but not for UPS and IPES. As typical valence exciton binding energies are in the range of 0.1–0.4 eV for molecular systems with strongly interacting π-systems,[189] the valence exciton binding energy cannot be neglected, but can also not entirely explain the observed deviations. Another possible contribution is the degradation of the Cu(I) complexes in UHV and binding of oxygen or water impurities in the glove box to the sample surface. Furthermore, charging could lead to the observation of an enlarged band gap and cannot be completely excluded even for the thin films. As the applicability of Tauc`s method to amorphous and microcrystalline samples of metal organic molecules has so far not been investigated comprehensively, no conclusions on which values for the HOMO-LUMO separations describe the actual separation better can be drawn.

So far, there are only few studies comparing different measurement techniques for the characterization of the HOMO and LUMO energies of metal organic emitter materials, which mostly only cover UPS and cyclic voltammetry as the most common methods. The herein presented study of Cu(I) NHetPHOS complexes **1**, **5**, and **6** by optical spectroscopy, cyclic voltammetry, PESA, as well as XPS, UPS, and IPES demonstrated that besides the chosen measurement technique, factors like sample preparation, sample type, and degradation throughout the sample preparation and measurement process can have a distinct impact on the observed HOMO and LUMO energies and HOMO-LUMO separations. Furthermore, a new approach for the assessment of the IP and EA of metal organic film samples by means of UPS and IPES along with XPS was established. The obtained data suggest that charging is not pronounced for the medium and thin film samples of the CuI NHetPHOS complexes, for which the IPs and EAs are in the range of 5.3–5.5 ($\pm$ 0.1) eV and 2.1–2.3 ($\pm$ 0.15) eV. Further measurements will be required to confirm this assumption and study the influence of the above mentioned factors and rule out degradation throughout the sample preparation and measurement process.

For the application of NHetPHOS complexes in efficient OLEDs with long device lifetimes, it is important to further determine the energy level lineup at the interfaces in an OLED architecture. Based on the herein presented approach, this could be conducted in a more detailed photoemission and inverse photoemission study of the HOMO and vacuum level position as well as the LUMO position as a function of the film thickness.

5   Summary

Luminescent Cu(I) NHetPHOS complexes are very efficient emitters, which have successfully been deployed in organic light-emitting diodes (OLEDs). They exhibit several advantages in comparison to other Cu(I) complexes: Apart from their high photoluminescence quantum yield, they cover the whole visible spectrum. Furthermore, they offer the possibility to be processed from solution due to good solubility, which allows for cost- and energy-efficient processing techniques such as printing rather than state-of-the-art vacuum processing.

In this work, a better overall understanding of the Cu(I) NHetPHOS complexes was developed by using suitable characterization techniques. The molecular structure of Cu(I) NHetPHOS complexes in non-crystalline form and upon solution-processing as well as the electronic structure with particular focus on the occupied and unoccupied molecular orbitals formed upon binding of the Cu- and N-atoms and the HOMO and LUMO energies of these materials were investigated. Furthermore, new red-emitting Cu(I) complexes were developed.

- Whereas there is a multiplicity of green- and yellow-emitting representatives, only few examples for red emitters have been reported. For the development of red-emitting Cu(I) complexes, quinoline, isoquinoline, and phenanthroline were identified as basic structures for N-aromatic ligands. The ligands were reacted with Cu(I) salts to yield Cu(I) complexes. The chemical and photophysical properties of these materials were characterized. Degradation of Cu(I) complexes with quinoline derivatives as ligands in the presence of water was observed. It was concluded that ligands that contain potential leaving groups may lead to degradation of the emissive complex and that strained or compressed bonds in the emissive complex may open up non-radiative decay pathways and again lead to disposition for degradation. Cu(I) complexes with alkyne-substituted ligands, which can undergo intramolecular-catalyzed, Cu(I)-catalyzed azide alkyne cycloadditions (CuAAc), were synthesized. These complexes can be deployed as crosslinkable emitters in OLEDs.

- The electronic structure of Cu(I) NHetPHOS complexes was investigated by means of resonant inelastic soft x-ray scattering and x-ray emission spectroscopy at the N K edge. Absorption and emission features in the N K RIXS map were assigned to occupied and unoccupied molecular orbitals formed upon binding of the N- and Cu-atoms, which have either dominating N- or Cu_2I_2-character. In combination with theoretical calculations, the N K core-excitonic state of the Cu(I) complexes was identified. Furthermore, these results provide first experimental evidence that the emissive transition is a metal-halide-to-ligand charge transfer. Theoretical calculations, which allow the assignment of specific molecular orbitals to absorption and emission features, are still ongoing.

- The HOMO and LUMO energies of the Cu(I) emitter are important parameters for OLED fabrication. The herein presented study of Cu(I) NHetPHOS complexes by optical spectroscopy, cyclic voltammetry, PESA, as well as XPS, UPS, and IPES demonstrated that besides the chosen measurement technique, factors like sample preparation, sample type, and degradation throughout the sample preparation and measurement process can have a distinct impact on the observed HOMO and LUMO energies. Furthermore, a new approach for the assessment of the IP and EA of metal organic film samples by means of UPS and IPES along with XPS was established.

- Furthermore, different methods for the characterization of the molecular structure of Cu(I) complexes – in particular, single crystal x-ray diffraction, photophysical measurements, IR and NMR spectroscopy, and x-ray absorption spectroscopy at the Cu K edge – were evaluated. X-ray absorption spectroscopy at the Cu K edge was used to characterize non-crystalline samples of Cu(I) NHetPHOS complexes. The structure determination of a NHetPHOS complex with a bridging bisphosphine ligand enabled the deployment of this material as emitter in OLEDs exhibiting a new quantum efficiency record for both solution- and vacuum-processed organic light emitting diodes with Cu(I) complexes as emitters. The analysis further suggested that the studied complexes retain their principle structure even when preparing thin films from solution and offer the opportunity to be used as emitters in solution-processed optoelectronic devices.

In summary, the understanding of the specific properties of the chemical, physical and photophysical properties of Cu(I)-OLED emitters was significantly improved. The findings from this study open the door to very efficient, solution-processed OLEDs with sustainable materials for the first time since the development of OLEDs started 25 years ago.

6   Experimental section

6.1   Analytical techniques

6.1.1   Chemical analysis

6.1.1.1 Nuclear magnetic resonance spectroscopy (NMR)

^{1}H (500 MHz) and ^{31}P (200 MHz) NMR spectra were recorded at room temperature on a Bruker Ultrashield 500 MHz instrument. Deuterated chloroform was purchased from Sigma-Aldrich and Deutero. Chemical shifts of all compounds are listed in parts per million (ppm) downfield from tetramethylsilane (TMS) and referenced to the residual protons or carbons of the solvent: 7.24 ppm (^{1}H) in chloroform-*d*. Chemical shifts in ^{31}P spectra were calculated without reference. Multiplicities of signals in ^{1}H NMR spectra are denoted as follows: s = singlet, bs = broad signal, q = quartet, quin = quintet, m = multiplet, dd = doublet of doublets, dt = doublet of triplet, tt = triplet of triplet. All spectra were analyzed according to first order splitting patterns. Coupling constants (*J*) are given in Hertz (Hz). For the analysis of the NMR data the software Mnova NMR Lite 5.2.5-5780 was used. The assignment of ^{1}H signals was done based on the prediction of ChemBioDraw Ultra 14.0.

6.1.1.2 Mass Spectrometry (EI-MS, FAB-MS)

Electron ionization (EI, 70 eV) and fast atom bombardment (FAB) mass spectra were measured on a Finnigan MAT 95 spectrometer. Molecular fragments and clusters are listed according to their mass to charge ratio (m/z). The protonated molecule ion (FAB) is referred to as and [M$^+$+H]. For FAB-MS the matrix 3-Nitrobenzylalcohol (3-NBA) was used. For Cu(I) complexes the following abbreviations were used: P = monodentate phosphorus ligand, N,P = bidentate N,P ligand, and N,N = bidentate N,N ligand.

6.1.1.3 Infrared Spectroscopy (IR)

IR spectra were recorded on a Bruker IFS 88 and Bruker Alpha. Solid samples were measured with ATR (attenuated total reflection) technique. Wavenumbers (ṽ) of the absorption bands are given in cm^{-1} and intensities are described as follows: vs = very strong (0–10% T), s = strong (11–40%), m = middle (41–70% T), w = weak (71–90% T) and vw = very weak (91–100% T).

6.1.1.4 Elemental analysis

Elemental analysis was performed on a system of Elementar (model: vario MICRO CUBE). A precision scale of Sartorius (model: M2P) was used. Values for carbon (C), hydrogen (H), and nitrogen (N) were stated as mass percent. The following abbreviations were used: calcd.: calculated theoretical value, found: value determined in the measurement.

6.1.2 Balance, purification, reaction control

6.1.2.1 Precision balance

Model: Mettler-Toledo, AB265-S/FACT.

6.1.2.2 Flash column chromatography

Flash column chromatography was carried out according to a procedure published by Still.[190] Silica gel was used as stationary phase and purchased from Merck (0.040 –0.063 mm). Sea sand from Merck (acid washed and calcined) was used as a protective layer on top of the stationary phase.

6.1.2.3 Centrifuge

Model: Eppendorf, centrifuge 5810R

6.1.3 Single crystal x-ray diffraction

Single crystal x-ray diffraction was measured at 123 K using Mo K$_\alpha$ radiation (λ = 0.71073 Å) on a Bruker-Nonius APEXII and Bruker-Nonius KappaCCD, or using Cu K$_\alpha$ radiation (λ = 1.54178 Å) on a Bruker D8 Venture Diffractometer with a Photon100 detector by Dr. Martin Nieger from the University of Helsinki. Direct (SHELXS-97)[191] or

Patterson methods were used to interpret the data, and refinement was carried out using SHELXL-97/2013/2014[192] (full-matrix least-squares on F_2). Non-hydrogen atoms were refined anisotropically, and hydrogen atoms were localized by difference electron density determination. Semi-absorption corrections were applied for all structures.

6.1.4 Photophysical measurements

6.1.4.1 UV/Vis absorption spectroscopy

UV/Vis spectra were recorded on a Thermo Scientific Evolution 201. Samples were measured as dichloromethane solutions (spectroscopic grade) at concentrations of 10^{-5} M, or as films on quartz substrates (spin-coated from 20 mg/ml dichloromethane solutions). The spectra were recalculated by:

$$A = \varepsilon \cdot c \cdot l \qquad (6.1)$$

(A: absorbance, ε: molar extinction coefficient, c: concentration, l: path length) and are displayed in extinction coefficients ε vs. wavelengths, if not otherwise noted.

6.1.4.2 Photoluminescence spectroscopy

Photoluminescence spectroscopy was done on spin-coated films (for details on the according procedure see 6.1.4.4), solid samples, or in solution. For solution measurements 2–3 mg of the sample was dissolved in 2 mL solvent (spectrophotometric grade). The cuvette was sealed with a rubber septum and the solution was flushed with nitrogen for 20 minutes. Solution samples were measured within one hour after preparation.

Photoluminescence spectra were recorded on a Horiba Scientific FluoroMax-4 with a 150 W xenon-arc lamp, excitation and emission monochromators and a Hamamatsu R928 photomultiplier tube. Emission and excitation spectra were corrected by standard calibration curves.

6.1.4.3 Photoluminescence quantum yields (PLQY)

Absolute photoluminescence quantum yields were measured using the integrating sphere method[193] on a Hamamatsu Photonics C9920-02G system. PLQYs are given as ratio of emitted photons to absorbed photons. Compounds were measured as powder or film samples with excitation at 350 nm. Calculation of the PLQYs and CIE coordinates was carried out with the software C9 920-02 V3.4.1. PLQY values within an accuracy of ±0.02.

6.1.4.4 Spin-coating

Samples for IPES, x-ray spectroscopy, and photophysical measurements were prepared on a spin-coater from SPS-Europe (model: SPIN150). The following program was used: 3 s at 400 rpm; 20 s at 1000 rpm, ramp rate = 1000 rpm/s; 10 s at 4000 rpm, ramp rate = 1000 rpm/s. Film samples were annealed for 10–30 min at 80 °C in air on a hot plate (IKA RCT basic).

Samples for XPS, UPS, and IPES measurements were prepared on a basic spin-coating device assembled from a computer ventilator with a peak spin speed of 3300 rpm. Film samples were not annealed but loaded into the load lock and kept at a pressure of 10^{-6} mbar for approximately 12 h prior to measurement.

6.1.4.5 Optical glass ware

Films for photophysical measurements were prepared on quartz glass or soda-lime glass substrates (round, diameter = 1 cm) by spin-coating (see 6.1.4.4). Solids and solution samples were measured in quartz glass cuvettes. All glass ware was cleaned by rinsing three times with dichloromethane, acetone, ethanol, and demineralized water. The glass ware was placed in 2% Hellmanex solution (Hellma Group) for 24 h and rinsed with demineralized water for three times. The glass ware was allowed to dry at room temperature in air.

6.1.5 Cyclic voltammetry

Cyclic voltammetry measurements in dichloromethane were recorded at room temperature under nitrogen flow with a system from CH Instruments (model 600E Series

Electrochemical Analyzer/Workstation). The measurements were performed with a glassy carbon working electrode, a platinum wire counter electrode and a platinum wire reference electrode at a scan rate of 100 mV/s. The supporting electrolyte was 0.1 M NBu_4PF_6. The measurement was calibrated against $FeCp_2/FeCp_2^+$.

6.1.6 Solid state NMR measurements[§§]

Neat film samples were obtained by spin-coating of concentrated solutions of the compounds on glass substrates (3 × 5 cm^2), drying, and scraping off the material. The samples were ground up to a fine powder and filled into 2.5 mm outer diameter rotors for solid state ^{31}P NMR measurements using magic angle spinning (MAS). To rule out changes in the samples by this sample preparation, the ground powders were observed under UV light (366 nm) and compared to the non-modified samples to confirm that the luminescence color did not change. All ^{31}P NMR experiments were performed at a resonance frequency of 242 MHz (corresponding to 600 MHz for ^1H) and 25 kHz sample spinning, using a wide-bore magnet and a Bruker Avance spectrometer (Bruker Biospin) equipped with a double-tuned MAS probe. The ^{31}P NMR spectra were acquired using ramped cross polarization (CP)[194] from protons, with a contact time of 5 ms and a radiofrequency field strength of 50 kHz to avoid long recycle times due to the long longitudinal relaxation time of ^{31}P in the copper complexes. Two-dimensional $^{31}P-^{31}P$ correlation experiments were obtained using a spin exchange pulse sequence,[195] deploying a mixing time of 200 ms and between 128-256 increments of ~40-80 µs dwell time each. The signal was acquired for 16–32 ms under 80 kHz proton decoupling. Typically 512 scans for the one-dimensional and 16–64 scans for the two-dimensional spectra were averaged, with a recycle delay of 3 s or 10 s in the one-dimensional spectra of NHetPHOS complex **3**. All spectra were processed using the Topspin software and applying a line broadening of 100 Hz.

[§§] Experimental details on the solid state NMR measurements were previously published in the context of this work: [102]   D. Volz, M. Wallesch, S. L. Grage, J. Göttlicher, R. Steininger, D. Batchelor, T. Vitova, A. S. Ulrich, C. Heske, L. Weinhardt, T. Baumann, S. Bräse, *Inorg. Chem.* **2014**, *53*, 7838-7847.

6.2   Synthesis and chemical analysis

6.2.1 General chemical working techniques

Starting materials and solvents were purchased from commercial suppliers (Sigma-Aldrich, VWR, TCI Europe, ABCR, and Roth) and used without further purification. Analytical grade solvents (ethyl acetate, cyclohexane, hexane, diethyl ether, dichloromethane, and methanol) for extraction, precipitation, and chromatography, and dry solvents (acetonitrile, dichloromethane, tetrahydrofuran, toluene, diethyl ether, hexane; < 50 mm H_2O) for reactions were also used as purchased without further purification. Solid reagents were used as powders and reactions carried out at room temperature if not otherwise noted. For low temperature reactions at –78 °C a dry ice/isopropanol cooling bath in a Dewar vessels were used. Conversion of the reactions was controlled by TLC, if applicable. Solvents were removed from the reaction mixture under reduced pressure by a rotary evaporator at 40 °C. Reactions with oxygen and/or water sensitive reagents were carried out under nitrogen atmosphere according the Schlenk technique.[196] Glassware was sealed with a septum and evacuated under heating and subsequently purged with nitrogen three times before usage. Solvents or liquids were transferred via syringes and V2A steel needles. Degassed solvents were obtained by purging the respective solvent with nitrogen.

6.2.2 Synthetic procedures and chemical analysis

6.2.2.1 Ligands

4-Methyl-2-(diphenylphosphino)-pyridin, MePyrPHOS. The synthesis of this compound was carried out according to a previously published protocol.[88] The ^{1}H-NMR and EI-MS data were in agreement with the data reported in the literature.[88]

4-Butyn-1-yl-(diphenylphosphino)-pyridin. butynylPyrPHOS. The synthesis of this compound was carried out according to a previously published protocol.[88] The [1] H-NMR and EI-MS were in agreement with the data reported in the literature.[88]

1-(diphenylphosphanyl)isoquinoline. IsoquinPHOS. This compound was kindly provided by Dr. Daniel Volz. The synthesis was carried out according to a previously published protocol.[197]

2-(But-3-yn-1-yl)-9-methyl-1,10-phenanthroline, (butynyl-Me-phen). To a solution of 1.41 mL (*i*-Pr)$_2$NH (1.02 g, 10.0 mmol, 0.7 equiv.) in THF (5 mL) was slowly added 4.02 mL of *n*-BuLi 2.5 M in hexanes (0.643 g, 10.0 mmol, 0.7 equiv.) at 0 °C. The mixture was stirred for 30 min at 0 °C before use. 2.92 mg 2,9-dimethylphenanthroline (14.0 mmol, 1.0 equiv.) was dissolved in anhydrous THF (5 mL) and. To this solution was added newly prepared LDA solution at 78 °C. The mixture was stirred at the same temperature for 1.5 h before 1.52 mL of a 80% solution of propargyl bromide in anhydrous toluene (1.62 mg, 13.6 mmol, 1.0 equiv.) was added. The mixture was stirred at 78 °C for 1 h and then at room temperature before quenching with water (5 mL). The slurry was extracted with EtOAc. The combined organic layers were dried over Na$_2$SO$_4$, and filtered. The solvent was removed in vacuum and the residual was purified by flash chromatography (silica, 5% triethylamine in 3:1 (v/v) cyclohexane/EtOAc 2,9-dimethylphenanthroline (R$_f$ = 0.18), product (R$_f$ = 0.21), disubstituted side product (R$_f$ = 0.45),) to yield the crude product, which was further deployed without further purification (yield 13%). HPLC*** (0.1% TFA in water/acetonitrile) yielded the pure product as TFA salt. HRMS and elemental analysis were done on the pure product. – ^{1}H NMR (500 MHz, chloroform-*d*) δ = 8.11–8.04 (m, 2 H), 7.63 (s, 2 H), 7.56–7.54 (m, 1 H), 7.43–7.42 (m, 1 H), 3.38 (t, *J* = 7.60 Hz, 2 H, C*H*$_2$CH$_2$CCH), 2.87 (s, 3 H, C*H*$_3$), 2.78 (dt, *J* = 2.46, 7.60, 2 H, CH$_2$C*H*$_2$CCH), 1.94 (t, *J* = 2.64 Hz, 1 H, CH$_2$CH$_2$CC*H*) ppm. – IR (ATR): 3052 (vw), 1578 (vw), 1436 (w), 1332 (vw), 1194 (w), 1173 (m), 1115(w), 1096 (w), 1072 (w), 1017 (vw), 840 (m), 793 (w), 754 (m), 723 (m), 696 (m), 551 (s), 513 (w), 498 (m), 447 (w), 406 (w) cm^{-1}. HRMS), *m/z*: (M$^+$ – C$_{17}$H$_{14}$N$_2$): calcd. 246.1157, found 246.1152. – [C$_{17}$H$_{14}$N$_2$ × C$_2$HF$_3$O$_2$] calcd. N 7.77, C 63.33, H 4.20, found N 7.59, C 62.92, H 4.11.

*** Purification by preparative HPLC was conducted by Karolin Kohnle.

(But-3-yn-1-yl)diphenylphosphite. butynylPHOS. This compound was kindly provided by Dr. Daniel Volz. The synthesis was carried out according to a previously published protocol.[89]

8-((Diphenylphosphanyl)oxy)quinoline. (POQ). 6.71 g (30.4 mmol, 1.0 equiv.) chlorodiphenylphosphine in 10 mL diethyl ether was slowly added to a solution of 4.41 g 8-hydroxyquinoline (30.4 mmol, 1.0 equiv.) in 40 mL diethyl ether and 8 mL Et$_3$N at 0 °C. The solution was stirred for 30 min. Precipitated Et$_3$NHCl was removed by filtration and washed with 10 mL diethyl ether. The solvent of the obtained organic phase was removed in vacuum, which yielded a yellow oil. The crude product was recrystallized from diethyl ether three times to yield colorless crystals (yield 28%). Upon exposure to traces of moisture or protic solvents, the product quickly decomposed to yield 8-hydroxyquinoline; therefore no elemental analysis matching the calculated values could be done. – ^{1}H NMR (500 MHz, chloroform-*d*) δ = 8.8 –8.83 (m, 1 H), 8.02–8.01 (m, 1 H), 7.71 (m, 4 H), 7.35–7.29 (m, 10H) ppm. – ^{31}P NMR (200 MHz, chloroform-*d*) δ = 117.0 ppm. – IR (ATR): 3052 (w), 1612 (vw), 1568 (w), 1496 (w), 1466 (m), 1432 (m), 1380 (w), 1310 (w), 1244 (m), 1165 (m), 1128 (w), 1084 (m), 1055 (m), 1024 (w), 997 (w), 914 (w), 826 (w), 809 (w), 792 (w), 776 (m), 739 (m), 691 (m), 555 (w), 543 (w), 522 (w), 495 (w), 458 (w), 422 (w) cm^{-1}. – FAB-MS (3-NBA), *m/z*: 387 [O(PPh$_2$)$_2$ + H$^+$], 330 [M + H$^+$], 329 [M$^+$], 201 [OPPh$_2$$^+$], 145 [8-hydroxyquinoline$^+$]. – calcd. N 4.25, C 76.59, H 4.90, found N 3.61, C 73.37, H 5.08.

6.2.2.2 Cu(I) complexes

Cu[phen(P(OEt)$_3$)$_2$]BF$_4$. Under nitrogen atmosphere, 252 mg tetrakis(acetonitrile)copper(I) tetrafluoroborate (0.800 mmol, 1.0 equiv.), 266 mg triethyl phosphite (1.60 mmol, 2.0 equiv.) and 144 mg phenanthroline (0.800 mmol, 1.0 equiv.) were put in a crimp vial equipped with a magnetic stir bar first and then dissolved in 12 mL MeCN. The reaction mixture was stirred until a clear solution was obtained. The volume of the solvent was reduced by means of a rotary evaporator until solid started to participate. The reaction mixture was placed in a fridge at –21 °C for one week. The yellow solid was collected by centrifugation of the reaction mixture. The product was dried overnight under reduced pressure (10^{-3} mbar) (yield 44%). – ^{1}H NMR (500 MHz, chloroform-*d*) δ = 9.15–9.14 (m, 2 H, phen-H), 8.66–8.65 (m, 2 H, phen-H), 8.09–8.05 (m, 4 H, phen-H), 3.79 (q, *J* = 7.0 Hz, 12 H, CH$_2$CH$_3$), 1.10 (t, *J* = 7.0 Hz, 18 H, CH$_2$CH$_3$) ppm. – ^{31}P NMR (200 MHz, chloroform-*d*) δ = 118.5 (bs) ppm. – FAB-MS (3-NBA)), *m/z:* 573 [(Cu$^+$)$_2$(phen)$_2$BF$_4^-$], 561 [Cu$^+$P(OEt)$_3$)$_3$], 545 [(Cu$^+$)$_2$P(OEt)$_3$)$_2$BF$_4^-$], 423 [Cu$^+$phen$_2$], 409 [Cu$^+$(phen)P(OEt)$_3$], 243 [Cu$^+$phen]. – calcd. N 4.23, C 43.49, H 5.78, found N 4.17, C 43.41, H 5.69. UV/VIS (CHCl$_3$): λ_{max} (ε in L mol^{-1} cm^{-1}) = 229 (3.629 10^4), 270 (2.907 10^4) nm.

CuI(butynyl-Me-phen)PPh$_3$. 95.2 mg CuI (0.500 mmol, 1.0 equiv.), 123 mg butynyl-Me-

phen (0.500 mmol, 1.0 equiv.) and 131 mg triphenylphosphine (0.500 mmol, 1.0 equiv.) were dissolved in dichloromethane. The solution was filtered through a syringe filter and the product was precipitated in hexane. The orange solid was stirred in diethyl ether for 12 h and dried under vacuum for 12 h. Single crystals suitable for single x-ray diffraction were obtained from a saturated solution of the product in acetonitrile (yield 56%). – ^{1}H NMR (500 MHz, chloroform-*d*) δ = 8.24–8.15 (m, 2 H), 7.76 (s, 2 H), 7.71–7.69 (m, 1 H), 7.47–7.39 (m, 1 H), 7.47–7.39 (m, 9H, PPh$_3$), 7.20–7.17 (m, 6 H, PPh$_3$), 3.44 (bs, 2 H, C*H$_2$*CH$_2$CCH), 2.85 (bs, 3 H, C*H$_3$*), 2.66 (bs, 2 H, CH$_2$C*H$_2$*CCH), 1.96 (bs, 1 H, CH$_2$CH$_2$CC*H*) ppm. – calcd. N 4.01, C 60.14, H 4.18, found N 3.82, C 59.75, H 3.99. – ^{31}P NMR (200 MHz, chloroform-*d*) δ = 29.5 ppm. – FAB-MS (3-NBA), *m/z:* 745 [Cu$_2$I(N,N)$_2^+$], 571 [Cu(N,N)PPh$_3^+$], 515 [Cu$_2$IPPh$_3^+$], 309 [Cu(N,N) $^+$]. – calcd. N 4.01, C 60.14, H 4.18, found N 3.72, C 60.38, H 4.04.

CuI(dmp)(butynylPHOS). 76.0 mg CuI (0.400 mmol, 1.0 equiv.), 83.0 mg 2,9-dimethylphenanthroline (0.400 mmol, 1.0 equiv.) and 102 mg triphenylphosphine (0.400 mmol, 1.0 equiv.) were dissolved in dichloromethane. The solution was filtered through a syringe filter and the product was precipitated in hexane. The solid was dried under vacuum (yield 49%). – ^{1}H NMR (500 MHz, chloroform-*d*) δ = 8.20 (d, *J* = 8.21 Hz, 2 H), 7.84–7.73 (m, 6 H), 7.77–7.50 (m, 2 H), 7.46–7.42 (m, 1 H), 7.35–7.25 (m, 5 H), 3.93 (bs, 2 H), 3.28 (bs, 1 H), 2.80 (bs, 2 H) ppm. – ^{31}P NMR (200 MHz, chloroform-*d*) δ = 32.1 ppm. – FAB-MS (3-NBA), *m/z:* 715 [Cu$_2$I(N,N)(P)$^+$], 669 [Cu$_2$I(N,N)$_2^+$], 525

[Cu(N,N)(P)⁺], 479 [Cu(N,N)₂⁺], 317 [Cu(P)⁺], 271 [Cu(N,N)⁺]. – calcd. N 4.19, C 55.74, H 4.53, found N 4.10, C 54.96, H 4.14.

Cu₂I₂(IsoquinPHOS)(butynylPHOS)₂. 76.0 mg CuI (0.400 mmol, 2.0 equiv.), 63 mg

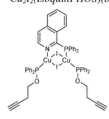

IsoquinPHOS (0.200 mmol, 1 equiv.), and 102 mg butynylPyrPHOS (0.400 mmol, 2.0 equiv.), were dissolved in anhydrous dichloromethane. The solution was filtered through a syringe filter and the product was precipitated in hexane. The solid was collected by centrifugation (yield 53%). – ¹H NMR (500 MHz, chloroform-*d*) δ = 9.42 (d, *J* = Hz, 1 H), 8.61 (d, *J* = Hz, 1 H), 7.02–7.29 (m, 34 H), 4.13–4.06 (m, 4 H), 3.15–3.10 (m, 1 H), 2.63–2.69 (m, 4 H) ppm. – ³¹P NMR (200 MHz, chloroform-*d*) δ = 28.5 ppm. – calcd. N 1.16, C 52.93, H 3.86, found N 0.95, C 49.95, H 3.43.

Cu₂Br₂(butynylPyrPHOS)(P(o-OMePh)₃)₂. 43.0 mg CuBr (0.300 mmol, 1.0 equiv.), 95.0

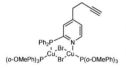

mg butynylPyrPHOS (0.300 mmol, 1.0 equiv.), and 211 mg tris(*o*-methoxyphenyl)phosphine (0.600 mmol, 2.0 equiv.) were dissolved in anhydrous dichloromethane. The solution was filtered through a syringe filter and the product was precipitated in hexane. The solid was collected by centrifugation (yield 51%). ¹H NMR (500 MHz, chloroform-*d*) δ = 8.49–8.43 (m, 1 H), 7.65–7.57 (m, 3 H), 7.38–7.32 (m, 3 H), 7.21–7.16 (m, 6 H), 6.86–6.81 (m, 19 H), 3.60 (bs, 18 H, OMe), 2.68 (bs, 2 H, C*H₂*CH₂CCH), 2.38 (bs, 2 H, CH₂C*H₂*CCH), 1.88 (bs, 1 H, CH₂CH₂CC*H*) ppm. – ³¹P NMR (200 MHz, chloroform-*d*) δ = 27.8, 20.9 ppm. – calcd. N 1.07, C 57.90, H 4.63, found N 0.72, C 58.04, H 4.71.

Cu₂I₂(butynylPyrPHOS)(PPh₃)₂. The synthesis of this compound was carried out according

to a previously published protocol.[89] The ¹H-NMR, FAB-MS, and the elemental analysis were in agreement with the data reported in the literature.[89] The product was obtained as powder from the reaction mixture by precipitation in hexane. Single crystals suitable for single crystal x-ray diffraction analysis were obtained from the supernatant after several days.

Cu₂Br₂(butynylPyrPHOS)₃. 39.6 mg CuCl (0.800 mml, 1.5 equiv.) and 252 mg

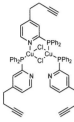

butynylPyrPHOS (0.400 mmol, 1.0 equiv.) were dissolved in anhydrous dichloromethane. The solution was filtered through a syringe filter and the product was precipitated in hexane. The solid was collected by centrifugation (yield 58%). ^{1}H NMR (500 MHz, chloroform-*d*) δ = 9.49–8.32 (m, 3 H), 7.60–6.90 (m, 36 H), 2.62 (bs, 6 H, C*H₂*CH₂CCH), 2.33 (bs, 6 H, CH₂C*H₂*CCH), 1.83 (bs, 3 H, CH₂CH₂CC*H*) ppm. – IR (ATR): 3287 (vw), 3045 (vw), 1586 (w), 1545 (vw), 1479(vw), 1433 (w), 1386 (w), 1182 (vw), 1093 (w), 1026 (vw), 988 (vw), 849 (vw), 741 (w), 691 (w), 632 (w), 497 (w), 464 (w), 437 (w). – calcd. N 3.67, C 66.14, H 4.76, found N 3.31, C 63.60, H 4.62. – FAB-MS (3-NBA), *m/z:* 891 [Cu₃Cl₂(N,P)₂$^+$], 791 [Cu₂Cl(N,P)₂$^+$], 693 [Cu((N,P)₂$^+$], 476 [Cu₂Cl(N,P)$^+$], 378 [Cu(N,P)$^+$]. – calcd. N 3.67, C 66.14, H 4.76, found N 3.31, C 63.60, H 44.62.

Cu₂Cl₂(MePyrPHOS)₃. The synthesis of this compound was carried out according to a

previously published protocol.[89] (Single crystals suitable for singe crystal x-ray diffraction were obtained by crystallization from a dichloromethane/hexane mixture and by diffusion method (acetonitrile/diethyl ether).

Cu₂I₂(POQ)₂. 95.2 mg CuI (0.500 mmol, 1 equiv.) and 165 mg *POQ* (0.500 mmol, 1 equiv.) were dissolved in 12 mL anhydrous dichloromethane by heating to the boiling point. The solution was filtered through a syringe filter and the product was precipitated in hexane. The yellow solid was collected by centrifugation and dried under vacuum for 1 h at 70 °C (yield 57%). Single crystals of Cu₂I₂(POQ)₂ suitable for single crystal x-ray diffraction were obtained by diffusion method (chloroform/cyclohexane). Upon exposure to traces of moisture or protic solvents, the product quickly decomposed; therefore no NMR spectra could be taken. – calcd. N 2.69, C 48.53, H 3.10, found N 2.34, C 47.82, H 2.94.

Cu$_2$Br$_2$(POQ)$_2$. 71.7 mg CuBr (0.500 mmol, 1 equiv.) and 165 mg POQ (0.500 mmol, 1 equiv.) were dissolved in 12 mL anhydrous dichloromethane by heating to the boiling point. The solution was filtered through a syringe filter and the product was precipitated in hexane. The pale yellow solid was collected by centrifugation and dried under vacuum for 1 h at 70 °C (yield 53%). Single crystals (pale yellow) of Cu$_2$Br$_2$(POQ)$_2$ suitable for single crystal x-ray diffraction were obtained by diffusion method (chloroform/cyclohexane). Upon exposure to traces of moisture or protic solvents, the product quickly decomposed; therefore no NMR spectra could be taken. – calcd. N 2.96, C 53.35, H 3.41, found N 2.53, C 51.47, H 3.27.

Cu$_4$Br$_4$(O(PPh$_2$)$_2$. Upon diffusion of diethyl ether into a solution of Cu$_2$Br$_2$(POQ)$_2$ in acetonitrile (crystallization by diffusion method (acetonitrile/diethyl ether)) colorless crystals of Cu$_4$Br$_4$(O(PPh$_2$)$_2$ suitable for single crystal x-ray diffraction were obtained.

Cu$_2$Br$_2$(O(PPh$_2$)$_2$)(PPh$_3$)$_2$. 43.0 mg CuBr (0.300 mmol, 1 equiv.), 98.8 mg POQ (0.300 mmol, 1 equiv.), and 78.7 mg triphenylphosphine (0.300 mmol, 1 equiv.) were dissolved in 12 mL anhydrous dichloromethane. The solution was filtered through a syringe filter and hexane was added. After one week yellow crystals suitable for single crystal x-ray diffraction were obtained. – calcd. N 0.00, C 60.16, H 4.21, found N 0.00, C 59.63, H 4.16.

Cu$_2$Cl$_2$(POQ)$_2$. 49.5 mg CuCl (0.500 mmol, 1 equiv.) and 165 mg POQ (0.500 mmol, 1 equiv.) were dissolved in 12 mL anhydrous dichloromethane by heating to the boiling point. The solution was filtered through a syringe filter and the product was precipitated in hexane. The pale yellow solid was collected by centrifugation and dried under vacuum for 1 h at 70 °C (yield 57%). Upon exposure to traces of moisture or protic solvents, the product quickly decomposed; therefore no NMR spectra could be taken. No single crystals suitable for single crystal x-ray diffraction could be obtained. – IR (ATR): 3057 (vw), 1582 (w), 1501 (w), 1468 (w), 1437 (w), 1378 (w), 1318 (w), 1270 (w), 1177 (w), 1112 (w), 1050 (w), 957 (w), 887 (w), 821 (w), 781 (w), 745 (w), 727 (w), 692 (w), 634 (w), 551 (w), 523 (w), 435 (w) cm^{-1}. – calcd. N 3.27, C 58.89, H 3.77, found N 2.99, C 58.63, H 3.78.

Cu$_2$(SCN)$_2$(POQ)$_2$. 60.82 mg CuSCN (0.500 mmol, 1 equiv.) and 165 mg POQ (0.500 mmol, 1 equiv.) were dissolved in 12 mL anhydrous dichloromethane by heating to the boiling point. The solution was filtered through a syringe filter and the product was precipitated in hexane. The pale yellow solid was collected by centrifugation and dried under vacuum for 1 h at 70 °C (yield 49%). Upon exposure to traces of moisture or protic solvents, the product quickly decomposed; therefore no NMR spectra could be taken. No single crystals suitable for single crystal x-ray diffraction could be obtained. – IR (ATR): 2114 (w), 1572 (vw), 1501 (vw), 1465 (w), 1435 (w), 1382 (w), 1310 (w), 1250 (w), 1172 (w), 1094 (w), 1063 (w), 1032 (vw), 905 (w), 828 (w), 814 (w), 794 (w), 783 (w), 756 (w), 740 (w), 711 (w), 696 (w), 588 (vw), 552 (w), 521 (w), 503 (w), 488 (w), 420 (w) cm^{-1}. – calcd. N 6.21, C 58.59, H 3.85, found N 5.89, C 58.58, H 3.84.

Cu$_4$(SCN)$_4$(O(PPh$_2$)$_2$. Crystals of Cu$_4$(SCN)$_4$(O(PPh$_2$)$_2$ $_2$ suitable for single crystal x-ray diffraction were obtained from a solution of Cu$_2$SCN$_2$(POQ)$_2$ in dichloromethane.

6.3   X-ray spectroscopic techniques

6.3.1 Sample preparation and setup for Cu K XAS measurements[†††]

Sample preparation – The chemical synthesis of the Cu(I) complexes was done according to procedures described in section 6.2. For EXAFS measurement samples were made by grinding and careful homogenization of 35 mg the material and 100 mg of cellulose to minimize self-absorption and pressing the mixture into pellets (diameter: 13 mm) using a laboratory press (Carver, max. 6 metric tons). Thin films were prepared via drop casting from a 1 M solution of the materials in MeCN onto kapton foil and dried at 60 °C for 20 min in air.

Measurement – The spectra were measured using the Si(111) double-crystal monochromator of the SUL-X wiggler beamline at the synchrotron radiation facility ANKA. The beam size at the sample position was approximately 800 µm × 800 µm. A thin Cu foil (4 µm) was measured simultaneously in transmission mode as a reference for energy calibration. Irradiation effects on spectral features were investigated by a series of quick scans, and the total exposure time on each sample spot was set such that irradiation effects in the spectra are negligible. The final scans were repeated on different sample positions to improve the signal-to-noise ratio. Up to three scans were accumulated for each spectrum. The energy step width in the XANES region was 0.4 eV. In the EXAFS region, the measuring time per k step (width 0.05 Å^{-1}) was modulated with the square root of k to improve the signal-to-noise ratio with increasing energy (increasing k). All measurements were performed in air and at room temperature. For normalization, the intensity of the primary beam was measured by an ionization chamber. Spectra were measured in transmission mode for pellet samples and fluorescence mode for film samples. Three ionization chambers (Oxford IC10 and ADC) were used to determine the primary beam intensity, as well as the sample and reference absorption. Fluorescence intensities were collected with a seven element Si(Li) solid state detector (SGX Sensortech, formerly

---

[†††] Experimental details on the sample preparation, setup for Cu K XAS measurements, and data analysis were previously published in the context of this work: [102]   D.   Volz,   M.   Wallesch,   S.   L.   Grage,   J. Göttlicher, R. Steininger, D. Batchelor, T. Vitova, A. S. Ulrich, C. Heske, L. Weinhardt, T. Baumann, S. Bräse, *Inorg. Chem.* **2014**, *53*, 7838-7847. [60]   D. Volz, Y. Chen, M. Wallesch, R. Liu, C. Fléchon, D. M. Zink, J. Friedrichs, H. Flügge, R. Steininger, J. Göttlicher, C. Heske, L. Weinhardt, S. Bräse, F. So, T. Baumann, *Adv. Mater.* **2015**, *17*, 2538–2543

Gresham) with the energy window set to the Cu Kα fluorescence emission lines. Signals were dead time-corrected, glitches were removed and signals were summed up for all channels, and divided by the input intensity. The energy scale was calibrated by assigning the first inflection point of the Cu foil spectrum to 8980.3 eV.

Analysis – The XAS data were analyzed using the program IFEFFIT, version 1.2.11.[198] The XANES data were processed using the AUTOBK algorithm of ATHENA.[199] A straight line was regressed to the data in the pre-edge range and a quadratic polynomial was used for the adaption of the post-edge background. The corrected data were normalized to an edge jump of unity. For EXAFS analysis, the spline function of ATHENA was used to extract the EXAFS function.

The EXAFS analysis was performed with the ARTEMIS[199] software. The scattering amplitudes and phases were calculated with the *ab initio* FEFF 8.4[200, 201] code. The Cu-N, Cu-P, Cu-I, and Cu-Cu single scattering paths were generated using interatomic distances from the single crystal x-ray diffraction analysis of crystalline samples of Cu(I) NHetPHOS complexes **1**, **2**, and **3**.

A k range of $k = 2.3–12.2$ Å^{-1}/$k = 2.3–11.2$ Å^{-1} was Fourier transformed (FT) for crystalline and powder/film samples. A shorter k range was chosen for film samples due to a higher signal-to-noise ratio. For the EXAFS fit analysis, a Hanning or Kaiser-Bessel window with $dk = 2$ was used. Fitting was performed in R space for 1.00–3.40 or 1.00–3.20 Å with multiple k-weights of 1, 2, and 3. The amplitude reduction factor $S_0^2 = 0.9$, which accounts for multi-electronic excitations, was obtained from a fit of an EXAFS spectrum of a metallic Cu foil measured in the same experimental conditions. S_0^2 was fixed to 0.9 during the fit. The assumption that S_0^2 is not dependent on the chemical properties of surrounding absorber atoms is a widely-used approximation.[116, 202-204] Note that $S_0^2 = 0.95$ is calculated with FEFF8.4 for the complexes studied here, but $S_0^2 = 0.9$ was maintained as an appropriate approximation to follow the customary approach. The threshold energy E_0 was allowed to vary for each fit but was constrained to the same value for all paths in a given fit. In a first step, the Cu-N, Cu-P and Cu-I paths were used to fit the first shell. The coordination numbers (N(N), N(P), N(I)), change in bond distance R (ΔR), and the mean-squared thermal and static atomic displacement (the Debye-Waller factor σ^2) were varied for all paths. In a second step, the second shell was included in the fit. The Cu-Cu path was added to the model and the coordination numbers and changes in distance were consecutively varied, such that the number of variables was kept half or less the number of independent data points. The N(I) coordination number was varied, whereas the N(Cu) was restrained (N(I) + N(Cu) = 3) during the fit. Independent variables were added by assigning one parameter for each ΔR and σ^2 to phosphorus/nitrogen, iodine, and copper. Best fits were chosen based on lowest values for χ_v^2 and R-factor.

6.3.2 Sample preparation and setup for N K XES and RIXS measurements

XES and RIXS spectra were recorded at beamline 8.0.1 of the Advanced Light Source (ALS), Lawrence Berkeley National Laboratory utilizing the high-transmission soft x-ray spectrometer[205] of the SALSA (solid and liquid spectroscopic analysis) endstation. A RIXS spectrum was collected at every point of a conventional absorption scan, resulting in RIXS maps that present the color-coded emission intensity as function of excitation and emission energies.[134] The excitation and emission energy axes were calibrated using reference measurements of N_2,[206] together with the elastically scattered Rayleigh line. To minimize the influence caused by molecules decomposed in the x-ray beam on the spectra, the sample was continuously scanned by with a scanning speed of $200\,\mu$m/s (corresponding to an exposure time of 0.15 s per sample spot). At this scanning speed, no evidence of decomposed molecules could be found in the spectra. The described materials were synthesized according to established protocols.[29, 88, 99] As reported for Cu(I) complexes **1** and **2**, the educts were dissolved in dichloromethane and the product was precipitated in hexane.[29, 99] The powder samples were ground and pressed into solid layers onto the sample holders.

6.3.3 Sample preparation and setup for XPS, UPES, and IPES measurements

Samples for XPS, UPS, and IPES measurements were prepared on a basic spin-coating device assembled from a computer ventilator with a peak spin speed of 3300 rpm. Films were spin-coated from toluene solutions on gold-coated silicon wafers as substrate. By dilution of the toluene solution thinner films were obtained. Film samples were not annealed but loaded into the load lock and kept at a pressure of 10^{-6} mbar for approximately 12 h prior to measurement.

UPS measurements were performed using He I excitation (He I at 21.2 eV) and for the XPS measurements a Mg K_α x-ray source (12 kV with 20 mA anode current) was used. The spectra were recorded with a PHOIBOS 150 electron analyzer with multi-channeltron detector. For the IPES experiments, a low-energy electron gun (STAIB) was used to irradiated the sample with low-energy electrons (between 7 and 15 eV). A Geiger-Müller type counter with SrF_2 window and Ar/I_2 filling was used as photon detector. All experiments were performed in ultrahigh vacuum with a base pressure below 1×10^{-9} mbar.

6.4   Single crystal x-ray diffraction analysis[‡‡‡]

6.4.1 Cu$_2$I$_2$(POQ)$_2$ (SB555)

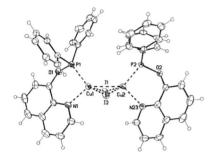

Empirical formula   C$_{42}$H$_{32}$Cu$_2$I$_2$N$_2$O$_2$P$_2$
Formula weight   1039.52
Temperature   123(2) K
Wavelength   0.71073 A
Crystal system, space group   Triclinic, P-1 (no.2)
Unit cell dimensions a = 11.928(1) A alpha = 86.55(1) deg.
b = 12.916(1) A beta = 83.21(1) deg.
c = 13.275(1) A gamma = 73.13(1) deg.
Volume   1942.7(3) A^3
Z, Calculated density 2, 1.777 Mg/m^3
Absorption coefficient   2.805 mm^-1
F(000) 1016
Crystal size   0.24 × 0.16 × 0.04 mm
Theta range for data collection   2.71 to 27.48 deg.
Limiting indices   -15<=h<=15, -16<=k<=16, -17<=l<=17
Reflections collected / unique   31683 / 8892 [R(int) = 0.0408]
Completeness to theta = 27.48   99.7%
Absorption correctionSemi-empirical from equivalents
Max. and min. transmission   0.9010 and 0.5673
Refinement method   Full-matrix least-squares on F^2
Data / restraints / parameters 8892 / 0 / 469
Goodness-of-fit on F^2   1.034
Final R indices [I>2sigma(I)]R1 = 0.0301, wR2 = 0.0677
R indices (all data)   R1 = 0.0446, wR2 = 0.0739
Largest diff. peak and hole   1.168 and -0.831 e.A^-3

[‡‡‡] The single crystal x-ray diffraction data reported in this work were obtained in collaboration with Dr. Martin Nieger (University of Helsinki). Single crystal were grown by the auther. The single crystal x-ray diffraction experiments and the analysis of the diffraction data was conducted by Dr. Martin Nieger.

6.4.2 $Cu_2Br_2(POQ)_2$ (SB563)

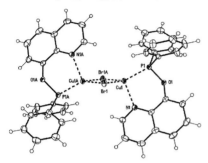

$C_{42}H_{32}Br_2Cu_2N_2O_2P_2$
$M_r = 945.53$
Monoclinic, $P2_1/n$
$a = 10.297 (2)$ Å
$b = 15.312 (1)$ Å
$c = 12.447 (2)$ Å
$\beta = 97.80 (1)°$
$V = 1944.3 (5)$ Å^3
$Z = 2$
Bruker-Nonius KappaCCD
diffractometer
Radiation source: fine.focus sealed tube
rotation in ϕ and ω, 2° scans
Absorption correction: multi-scan
SADABS (Shelxdrick, 2008)
$T_{min} = 0.605$, $T_{max} = 0.725$
25394 measured reflections
4449 independent reflections
Refinement on F^2

Least-squares matrix: full

$R[F^2 > 2\sigma(F^2)] = 0.033$

$wR(F^2) = 0.066$
$S = 1.09$

4449 reflections
235 parameters
0 restraints

$F(000) = 944$
$D_x = 1.615$ Mg m^{-3}
Mo $K\alpha$ radiation, $\lambda = 0.71073$ Å
Cell parameters from 204 reflections
$\theta = 2.5–25.0°$
$\mu = 3.27$ mm^{-1}
$T = 123$ K
Blocks, pale yellow
$0.20 \times 0.15 \times 0.10$ mm
3556 reflections with $I > 2\sigma(I)$

$R_{int} = 0.044$
$\theta_{max} = 27.5°$, $\theta_{min} = 2.7°$
$h = -13 \rightarrow 13$

$k = -19 \rightarrow 19$
$l = -16 \rightarrow 16$

Primary atom site location: structure-invariant direct methods
Secondary atom site location: difference Fourier map
Hydrogen site location: inferred from neighboring sites
H-atom parameters constrained
$w = 1/[\sigma^2(F_o^2) + (0.021P)^2 + 2.4027P]$
where $P = (F_o^2 + 2F_c^2)/3$
$(\Delta/\sigma)_{max} = 0.001$
$\Delta\rangle_{max} = 0.49$ e Å^{-3}
$\Delta\rangle_{min} = -0.38$ e Å^{-3}

6.4.3 $Cu_4Br_4(O(PPh_2)_2$ (SB562)

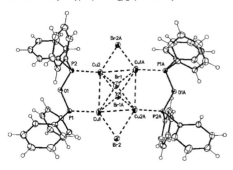

C$_{48}$H$_{40}$Br$_4$Cu$_4$O$_2$P$_4$
$M_r = 1346.48$
Monoclinic, $P2_1/n$
$a = 10.685$ (1) Å
$b = 16.416$ (2) Å
$c = 13.617$ (2) Å
$\beta = 92.27$ (1)°
$V = 2386.6$ (5) Å^3
$Z = 2$
Bruker-Nonius KappaCCD
diffractometer
Radiation source: fine-focus sealed tube
rotation in ϕ and ω, 2° scans
Absorption correction: multi-scan
SADABS (Sheldrick, 2008)
$T_{min} = 0.291$, $T_{max} = 0.471$
47400 measured reflections
5469 independent reflections
Refinement on F^2

Least-squares matrix: full

$R[F^2 > 2\sigma(F^2)] = 0.022$

$wR(F^2) = 0.047$
$S = 1.06$

5469 reflections
280 parameters
0 restraints

$F(000) = 1320$
$D_x = 1.874$ Mg m^{-3}
Mo $K\alpha$ radiation, $\lambda = 0.71073$ Å
Cell parameters from 371 reflections
$\theta = 2.5–25.0°$
$\mu = 5.28$ mm^{-1}
$T = 123$ K
Blocks, colorless
$0.35 \times 0.25 \times 0.20$ mm
4740 reflections with $I > 2\sigma(I)$

$R_{int} = 0.027$
$\theta_{max} = 27.5°$, $\theta_{min} = 2.7°$
$h = -13 \rightarrow 13$

$k = -21 \rightarrow 21$
$l = -17 \rightarrow 17$

Primary atom site location: structure-invariant direct methods
Secondary atom site location: difference Fourier map
Hydrogen site location: inferred from neighboring sites
H-atom parameters constrained
$w = 1/[\sigma^2(F_o^2) + (0.0138P)^2 + 3.4792P]$
where $P = (F_o^2 + 2F_c^2)/3$
$(\Delta/\sigma)_{max} = 0.002$
$\Delta\rangle_{max} = 1.14$ e Å^{-3}
$\Delta\rangle_{min} = -0.93$ e Å^{-3}

6.4.4 $Cu_2Br_2(O(PPh_2)_2)(PPh_3)_2$ (SB705)

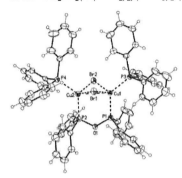

Cu-atoms disordered (96:4)
$C_{60}H_{50}Br_2Cu_2OP_4$
$M_r = 1197.78$
Triclinic, P-1 (no.2)
$a = 10.321$ (2) Å
$b = 14.058$ (1) Å
$c = 19.264$ (2) Å
$\alpha = 95.40$ (1)°
$\beta = 104.81$ (1)°
$\gamma = 100.93$ (1)°
$V = 2623.1$ (6) Å^3
Bruker-Nonius KappaCCD diffractometer
Radiation source: fine-focus sealed tube
rotation in ϕ and ω, 1° scans
Absorption correction: multi-scan
$SADABS$ (Sheldrick, 2008)
$T_{min} = 0.717$, $T_{max} = 0.862$
44354 measured reflections
12037 independent reflections
Refinement on F^2

Least-squares matrix: full

$R[F^2 > 2\sigma(F^2)] = 0.037$

$wR(F^2) = 0.086$
$S = 1.04$

12037 reflections
630 parameters
0 restraints

$Z = 2$
$F(000) = 1212$
$D_x = 1.516$ Mg m^{-3}
Mo $K\alpha$ radiation, $\lambda = 0.71073$ Å
Cell parameters from 163 reflections
$\theta = 2.5–25.0°$
$\mu = 2.50$ mm^{-1}
$T = 123$ K
Blocks, yellow
$0.20 \times 0.20 \times 0.06$ mm
9539 reflections with $I > 2\sigma(I)$

$R_{int} = 0.037$
$\theta_{max} = 27.5°$, $\theta_{min} = 2.6°$
$h = -13 \rightarrow 13$

$k = -18 \rightarrow 18$
$l = -25 \rightarrow 25$

Primary atom site location: heavy-atom method
Secondary atom site location: difference Fourier map
Hydrogen site location: inferred from neighboring sites
H-atom parameters constrained
$w = 1/[\sigma^2(F_o^2) + (0.0394P)^2 + 1.850P]$
where $P = (F_o^2 + 2F_c^2)/3$
$(\Delta/\sigma)_{max} = 0.001$
$\Delta\rangle_{max} = 0.92$ e Å^{-3}
$\Delta\rangle_{min} = -0.58$ e Å^{-3}

6.4.5 Cu₂(SCN)₂(POQ)₂ (SB702)

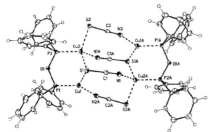

C$_{52}$H$_{40}$Cu$_4$N$_4$O$_2$P$_4$S$_4$
M_r = 1259.16
Monoclinic, $P2_1/n$ (no.14)
a = 11.665 (1) Å
b = 17.923 (2) Å
c = 12.097 (1) Å
β = 94.35 (1)°
V = 2521.9 (4) Å^3
Z = 2
Bruker-Nonius KappaCCD diffractometer
Radiation source: fine-focus sealed tube
rotation in ϕ and ω, 2° scans
Absorption correction: multi-scan
$SADABS$ (Sheldrick, 2008)
T_{min} = 0.698, T_{max} = 0.862
40990 measured reflections
5785 independent reflections
Refinement on F^2

Least-squares matrix: full

$R[F^2 > 2\sigma(F^2)]$ = 0.026

$wR(F^2)$ = 0.059
S = 1.06

5785 reflections
316 parameters
0 restraints

$F(000)$ = 1272
D_x = 1.658 Mg m^{-3}
Mo $K\alpha$ radiation, λ = 0.71073 Å
Cell parameters from 172 reflections
θ = 2.5–25.0°
μ = 2.00 mm^{-1}
T = 123 K
Plates, yellow
0.30 × 0.12 × 0.06 mm
4959 reflections with $I > 2\sigma(I)$

R_{int} = 0.036
θ_{max} = 27.5°, θ_{min} = 2.6°
h = -15→15

k = -23→23
l = -15→15

Primary atom site location: structure-invariant direct methods
Secondary atom site location: difference Fourier map
Hydrogen site location: difference Fourier map
H-atom parameters constrained
$w = 1/[\sigma^2(F_o^2) + (0.0242P)^2 + 1.8372P]$
where $P = (F_o^2 + 2F_c^2)/3$
$(\Delta/\sigma)_{max}$ = 0.002
$\Delta\rangle_{max}$ = 0.45 e Å^{-3}
$\Delta\rangle_{min}$ = -0.31 e Å^{-3}

6.4.6 CuI(butynyl-Me-phen)PPh₃ (SB590)

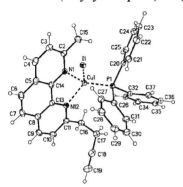

C₃₅H₂₉CuIN₂P
$M_r = 699.01$
Triclinic, P-1 (no.2)
$a = 10.030$ (1) Å
$b = 12.441$ (1) Å
$c = 13.157$ (1) Å
$\alpha = 82.74$ (1)°
$\beta = 71.68$ (1)°
$\gamma = 71.09$ (1)°
$V = 1473.9$ (3) Å³
Bruker-Nonius KappaCCD
diffractometer
Radiation source: fine-focus sealed tube
rotation in ϕ and ω, 2° scans
Absorption correction: multi-scan
SADABS (Sheldrick, 2008)
$T_{min} = 0.728$, $T_{max} = 0.801$
36427 measured reflections
6732 independent reflections
Refinement on F^2

Least-squares matrix: full

$R[F^2 > 2\sigma(F^2)] = 0.017$

$wR(F^2) = 0.043$
$S = 1.08$

6732 reflections
362 parameters
0 restraints

$Z = 2$
$F(000) = 700$
$D_x = 1.575$ Mg m⁻³
Mo $K\alpha$ radiation, $\lambda = 0.71073$ Å
Cell parameters from 269 reflections
$\theta = 2.5–25.0°$
$\mu = 1.87$ mm⁻¹
$T = 123$ K
Blocks, orange
$0.30 \times 0.15 \times 0.10$ mm
6257 reflections with $I > 2\sigma(I)$

$R_{int} = 0.021$
$\theta_{max} = 27.5°$, $\theta_{min} = 3.1°$
$h = -13 \rightarrow 13$

$k = -16 \rightarrow 16$
$l = -17 \rightarrow 17$

Primary atom site location: structure-invariant direct methods
Secondary atom site location: difference Fourier map
Hydrogen site location: difference Fourier map
H-atom parameters constrained
$w = 1/[\sigma^2(F_o^2) + (0.0172P)^2 + 0.8724P]$
where $P = (F_o^2 + 2F_c^2)/3$
$(\Delta/\sigma)_{max} = 0.003$
$\Delta\rangle_{max} = 0.39$ e Å⁻³
$\Delta\rangle_{min} = -0.40$ e Å⁻³

6.4.7 Cu₂Cl₂(MePyrPHOS)₃·Toluene (SB703)

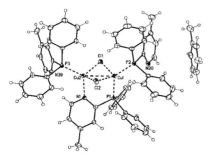

$C_{54}H_{48}Cl_2Cu_2N_3P_3 \cdot C_7H_8$
$M_r = 1121.97$
Monoclinic, *C2/c (no.15)*
$a = 43.420 (4)$ Å
$b = 11.327 (1)$ Å
$c = 23.710 (2)$ Å
$\beta = 113.06 (1)°$
$V = 10729.2 (18)$ Å^3
$Z = 8$
Bruker-Nonius KappaCCD diffractometer
Radiation source: fine-focus sealed tube
rotation in ϕ and ω, 1° scans
Absorption correction: multi-scan
SADABS (Shelxrick, 2008)
$T_{min} = 0.862$, $T_{max} = 0.962$
63184 measured reflections
12319 independent reflections
Refinement on F^2

Least-squares matrix: full

$R[F^2 > 2\sigma(F^2)] = 0.045$

$wR(F^2) = 0.088$
$S = 1.02$

12319 reflections
644 parameters
0 restraints

$F(000) = 4640$
$D_x = 1.389$ Mg m^{-3}
Mo $K\alpha$ radiation, $\lambda = 0.71073$ Å
Cell parameters from 138 reflections
$\theta = 2.5–25.0°$
$\mu = 1.02$ mm^{-1}
$T = 123$ K
Plates, yellow
$0.20 \times 0.10 \times 0.04$ mm
8862 reflections with $I > 2\sigma(I)$

$R_{int} = 0.064$
$\theta_{max} = 27.5°$, $\theta_{min} = 2.6°$
$h = -56 \rightarrow 56$

$k = -14 \rightarrow 14$
$l = -30 \rightarrow 30$

Primary atom site location: structure-invariant direct methods
Secondary atom site location: difference Fourier map
Hydrogen site location: difference Fourier map
H-atom parameters constrained
$w = 1/[\sigma^2(F_o^2) + (0.0291P)^2 + 21.550P]$
where $P = (F_o^2 + 2F_c^2)/3$
$(\Delta/\sigma)_{max} = 0.001$
$\Delta\rangle_{max} = 0.50$ e Å^{-3}
$\Delta\rangle_{min} = -0.39$ e Å^{-3}

6.4.8 $Cu_2Cl_2(MePyrPHOS)_3 \cdot 0.5$ Dietylether·Acetonitrile (SB704)

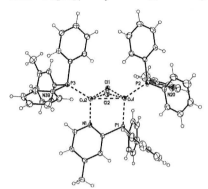

$C_{54}H_{48}Cl_2Cu_2N_3P_3 \cdot 0.5(C_4H_{10}O) \cdot C_2H_3N$
$M_r = 1107.95$
Monoclinic, $C2/c$ (no.15)
$a = 46.551$ (4) Å
$b = 11.335$ (1) Å
$c = 21.616$ (2) Å
$\beta = 110.73$ (1)°
$V = 10667.4$ (18) Å^3
$Z = 8$
Bruker-Nonius KappaCCD
diffractometer
Radiation source: fine-focus sealed tube
rotation in ϕ and ω, 1° scans
Absorption correction: multi-scan
SADABS (Sheldrick, 2008)
$T_{min} = 0.642$, $T_{max} = 0.746$
41885 measured reflections
12231 independent reflections
Refinement on F^2

Least-squares matrix: full

$R[F^2 > 2\sigma(F^2)] = 0.027$

$wR(F^2) = 0.067$
$S = 1.05$

12231 reflections
628 parameters
20 restraints

$F(000) = 4584$
$D_x = 1.380$ Mg m^{-3}
Mo $K\alpha$ radiation, $\lambda = 0.71073$ Å
Cell parameters from 217 reflections
$\theta = 2.5$–25.0°
$\mu = 1.03$ mm^{-1}
$T = 123$ K
Blocks, yellow
$0.50 \times 0.35 \times 0.25$ mm
10607 reflections with $I > 2\sigma(I)$

$R_{int} = 0.027$
$\theta_{max} = 27.5°$, $\theta_{min} = 2.5°$
$h = -59 \rightarrow 60$
$k = -14 \rightarrow 14$
$l = -28 \rightarrow 27$

Primary atom site location: structure-
invariant direct methods
Secondary atom site location: difference
Fourier map
Hydrogen site location: inferred from
neighboring sites
H-atom parameters constrained
$w = 1/[\sigma^2(F_o^2) + (0.024P)^2 + 16.180P]$
where $P = (F_o^2 + 2F_c^2)/3$
$(\Delta/\sigma)_{max} = 0.002$
$\Delta\rangle_{max} = 0.47$ e Å^{-3}
$\Delta\rangle_{min} = -0.47$ e Å^{-3}

6.4.9 Cu₂I₂(butynylPyrPHOS)(PPh₃)₂ (SB447)

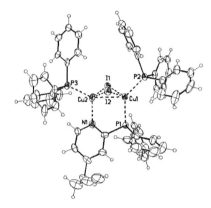

Empirical formula $C_{57}H_{48}Cu_2I_2NP_3$
Formula weight   1220.75
Temperature   123(2) K
Wavelength   0.71073 A
Crystal system, space group Monoclinic, P2(1)/n (no.14)
Unit cell dimensions a = 12.044(2) A alpha = 90 deg.
b = 24.064(4) A beta = 105.93(2) deg.
c = 18.501(2) A gamma = 90 deg.
Volume   5156.2(13) A^3
Z, Calculated density 4, 1.573 Mg/m^3
Absorption coefficient   2.153 mm^-1
F(000) 2424
Crystal size   0.16 × 0.08 × 0.02 mm
Theta range for data collection   3.00 to 25.03 deg.
Limiting indices   -14<=h<=14, -24<=k<=28, -22<=l<=17
Reflections collected / unique   26411 / 9070 [R(int) = 0.0916]
Completeness to theta = 25.03   99.5%
Absorption correctionSemi-empirical from equivalents
Max. and min. transmission 0.9561 and 0.8618
Refinement method   Full-matrix least-squares on F^2
Data / restraints / parameters 9070 / 0 / 586
Goodness-of-fit on F^2   1.005
Final R indices [I>2sigma(I)]R1 = 0.0580, wR2 = 0.0832
R indices (all data)   R1 = 0.1290, wR2 = 0.0995
Largest diff. peak and hole   1.113 and -0.846 e.A^-3

6.4.10Cu₂I₄(butynylPyrPHOS)₂ (SB485)

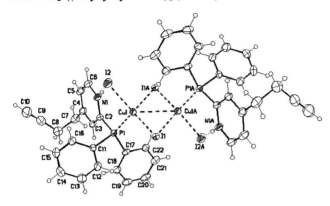

Empirical formula $C_{42}H_{38}Cu_2I_4N_2P_2$
Formula weight   1267.36
Temperature   120(2) K
Wavelength   0.71073 A
Crystal system, space group   Monoclinic, P2(1)/n (no.14)
Unit cell dimensions   a = 9.7873(3) A alpha = 90 deg.
        b = 13.8982(4) A beta = 91.880(1) deg.
c = 15.7975(4) A gamma = 90 deg.
Volume   2147.71(11) A^3
Z, Calculated density 2, 1.960 Mg/m^3
Absorption coefficient   3.971 mm^-1
F(000) 1208
Crystal size   0.24 × 0.16 × 0.08 mm
Theta range for data collection   2.97 to 28.28 deg.
Limiting indices   -13<=h<=13, -18<=k<=17, -21<=l<=20
Reflections collected / unique   14099 / 5301 [R(int) = 0.0292]
Completeness to theta = 28.28   99.4%
Absorption correctionSemi-empirical from equivalents
Max. and min. transmission 0.7457 and 0.6137
Refinement method   Full-matrix least-squares on F^2
Data / restraints / parameters 5301 / 1 / 238
Goodness-of-fit on F^2   1.065
Final R indices [I>2sigma(I)]R1 = 0.0396, wR2 = 0.1141
R indices (all data)   R1 = 0.0471, wR2 = 0.1197
Extinction coefficient SHELXL-97 .Sheldrick, 1997)
Largest diff. peak and hole   1.784 and -2.239 e.A^-3

6.4.11 CuBr(dmp)(P(OEt)₃) (SB591)

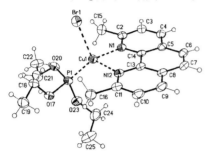

$C_{20}H_{27}BrCuN_2O_3P$
$M_r = 517.85$
Triclinic, $P\bar{1}$
$a = 9.295\ (1)$ Å
$b = 10.519\ (1)$ Å
$c = 11.913\ (1)$ Å
$\alpha = 98.34\ (1)°$
$\beta = 94.86\ (1)°$
$\gamma = 99.01\ (1)°$
$V = 1131.34\ (19)$ Å³
Bruker-Nonius KappaCCD
diffractometer
Radiation source: fine-focus sealed tube
rotation in ϕ and ω, 2° scans
Absorption correction: multi-scan
SADABS (Sheldrick, 2008)
$T_{min} = 0.650$, $T_{max} = 0.862$
18342 measured reflections
5161 independent reflections
Refinement on F^2

Least-squares matrix: full

$R[F^2 > 2\sigma(F^2)] = 0.047$

$wR(F^2) = 0.118$
$S = 1.07$

5161 reflections
255 parameters
0 restraints

$Z = 2$
$F(000) = 528$
$D_x = 1.520$ Mg m⁻³
Mo $K\alpha$ radiation, $\lambda = 0.71073$ Å
Cell parameters from 160 reflections
$\theta = 2.5–25.0°$
$\mu = 2.82$ mm⁻¹
$T = 123$ K
Blocks, yellow
$0.30 \times 0.10 \times 0.06$ mm
4207 reflections with $I > 2\sigma(I)$

$R_{int} = 0.044$
$\theta_{max} = 27.5°$, $\theta_{min} = 2.7°$
$h = -12 \rightarrow 12$

$k = -13 \rightarrow 13$
$l = -15 \rightarrow 15$

Primary atom site location: heavy-atom
method
Secondary atom site location: difference
Fourier map
Hydrogen site location: difference Fourier
map
H-atom parameters constrained
$w = 1/[\sigma^2(F_o^2) + (0.040P)^2 + 3.3893P]$
where $P = (F_o^2 + 2F_c^2)/3$
$(\Delta/\sigma)_{max} = 0.001$
$\Delta\rangle_{max} = 0.66$ e Å⁻³
$\Delta\rangle_{min} = -0.98$ e Å⁻³

6.4.12 $Cu_2Br_2(PPh_3)_3$ (SB557) [§§§]

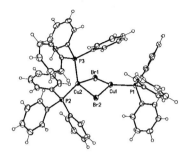

$C_{54}H_{45}Br_2Cu_2P_3$
$M_r = 1073.71$
Monoclinic, $P2_1/n$
$a = 19.070 (2)$ Å
$b = 9.880 (1)$ Å
$c = 26.225 (3)$ Å
$\beta = 109.82 (1)°$
$V = 4648.4 (9)$ Å^3
$Z = 4$
Bruker-Nonius KappaCCD
diffractometer
Radiation source: fine-focus sealed tube
rotation in ϕ and ω, 1° scans
Absorption correction: multi-scan
SADABS (Sheldrick 2008)
$T_{min} = 0.626$, $T_{max} = 0.862$
65128 measured reflections
10644 independent reflections
Refinement on F^2

Least-squares matrix: full

$R[F^2 > 2\sigma(F^2)] = 0.027$

$wR(F^2) = 0.056$
$S = 1.05$

10644 reflections
550 parameters
0 restraints

$F(000) = 2168$
$D_x = 1.534$ Mg m^{-3}
Mo $K\alpha$ radiation, $\lambda = 0.71073$ Å
Cell parameters from 324 reflections
$\theta = 2.5–25°$
$\mu = 2.77$ mm^{-1}
$T = 123$ K
Rods, colorless
$0.24 \times 0.12 \times 0.03$ mm
8644 reflections with $I > 2\sigma(I)$

$R_{int} = 0.039$
$\theta_{max} = 27.5°$, $\theta_{min} = 2.6°$
$h = -24\rightarrow24$

$k = -12\rightarrow12$
$l = -34\rightarrow34$

Primary atom site location: structure-invariant direct methods
Secondary atom site location: difference Fourier map
Hydrogen site location: inferred from neighboring sites
H-atom parameters constrained
$w = 1/[\sigma^2(F_o^2) + (0.0158P)^2 + 4.311P]$
where $P = (F_o^2 + 2F_c^2)/3$
$(\Delta/\sigma)_{max} = 0.002$
$\Delta\rangle_{max} = 0.41$ e Å^{-3}
$\Delta\rangle_{min} = -0.34$ e Å^{-3}

[§§§] This crystal structure was previously published, e.g., J. Dyason, L. Engelhardt, C. Pakawatchai, P. Healy, A. White, *Aust. J. Chem.* **1985**, *38*, 1243-1250.

6.4.13 CuBr(PPh$_3$)$_2$ (SB558)****

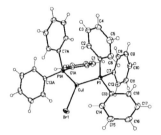

C$_{36}$H$_{30}$BrCuP$_2$
$M_r = 667.99$
Monoclinic, $C2/c$
$a = 24.604\ (2)$ Å
$b = 9.024\ (1)$ Å
$c = 15.052\ (1)$ Å
$\beta = 116.96\ (1)°$
$V = 2978.8\ (5)$ Å^3
$Z = 4$
Bruker-Nonius KappaCCD diffractometer
Radiation source: fine-focus sealed tube
rotation in ϕ and ω, 2° scans
Absorption correction: multi-scan
SADABS (Sheldrick, 2008)
$T_{min} = 0.692$, $T_{max} = 0.837$
29699 measured reflections
3407 independent reflections
Refinement on F^2

Least-squares matrix: full

$R[F^2 > 2\sigma(F^2)] = 0.019$

$wR(F^2) = 0.050$
$S = 1.05$

3407 reflections
182 parameters
0 restraints

$F(000) = 1360$
$D_x = 1.490$ Mg m^{-3}
Mo $K\alpha$ radiation, $\lambda = 0.71073$ Å
Cell parameters from 178 reflections
$\theta = 2.5$–$25.0°$
$\mu = 2.21$ mm^{-1}
$T = 123$ K
Blocks, colorless
$0.48 \times 0.16 \times 0.08$ mm
3151 reflections with $I > 2\sigma(I)$

$R_{int} = 0.022$
$\theta_{max} = 27.5°$, $\theta_{min} = 2.6°$
$h = -31 \rightarrow 31$

$k = -11 \rightarrow 11$
$l = -19 \rightarrow 19$

Primary atom site location: structure-invariant direct methods
Secondary atom site location: difference Fourier map
Hydrogen site location: inferred from neighboring sites
H-atom parameters constrained
$w = 1/[\sigma^2(F_o^2) + (0.024P)^2 + 3.5027P]$
where $P = (F_o^2 + 2F_c^2)/3$
$(\Delta/\sigma)_{max} = 0.001$
$\Delta\rangle_{max} = 0.36$ e Å^{-3}
$\Delta\rangle_{min} = -0.39$ e Å^{-3}

**** Polymorphs of this crystal structure were previously published, e.g., P. H. Davis, R. L. Belford, I. C. Paul, *Inorg. Chem.* **1973**, *12*, 213-218.

7 References

[1] G. Chansin, K. Ghaffarzadeh, H. Zervos, *OLED Display Forecast 2015-2025: The Rise of Plastic and Flexible Displays*, http://www.idtechex.com/research/reports/oled-display-forecast-2015-2025-the-rise-of-plastic-and-flexible-displays-000426.asp?viewopt=showall, **2015**, (date accessed: 22.09.2015).

[2] N. Koch, *ChemPhysChem* **2007**, *8*, 1438-1455.

[3] H. Ishii, K. Sugiyama, E. Ito, K. Seki, *Adv. Mater.* **1999**, *11*, 605-625.

[4] H. Ishii, K. Seki, *IEEE Trans. Electron Devices* **1997**, *44*, 1295-1301.

[5] M. T. Greiner, M. G. Helander, W. M. Tang, Z. B. Wang, J. Qiu, Z. H. Lu, *Nat. Mater.* **2012**, *11*, 76-81.

[6] H. Wang, P. Amsalem, G. Heimel, I. Salzmann, N. Koch, M. Oehzelt, *Adv. Mater.* **2014**, *26*, 925-930.

[7] L. Ley, Y. Smets, C. I. Pakes, J. Ristein, *Adv. Funct. Mater.* **2013**, *23*, 794-805.

[8] I. Lange, J. C. Blakesley, J. Frisch, A. Vollmer, N. Koch, D. Neher, *Phys. Rev. Lett.* **2011**, *106*, 216402.

[9] J. C. Blakesley, N. C. Greenham, *J. Appl. Phys.* **2009**, *106*, 034507.

[10] S. Braun, W. R. Salaneck, M. Fahlman, *Adv. Mater.* **2009**, *21*, 1450-1472.

[11] H. Fukagawa, S. Kera, T. Kataoka, S. Hosoumi, Y. Watanabe, K. Kudo, N. Ueno, *Adv. Mater.* **2007**, *19*, 665-668.

[12] N. Koch, A. Elschner, J. P. Rabe, R. L. Johnson, *Adv. Mater.* **2005**, *17*, 330-335.

[13] M. Knupfer, H. Peisert, *Phys. Status Solidi A* **2004**, *201*, 1055-1074.

[14] G. Koller, B. Winter, M. Oehzelt, J. Ivanco, F. P. Netzer, M. G. Ramsey, *Org. Electron.* **2007**, *8*, 63-68.

[15] H. Vázquez, F. Flores, R. Oszwaldowski, J. Ortega, R. Pérez, A. Kahn, *Appl. Surf. Sci.* **2004**, *234*, 107-112.

[16] S. Winkler, J. Frisch, R. Schlesinger, M. Oehzelt, R. Rieger, J. Räder, J. P. Rabe, K. Müllen, N. Koch, *J. Phys. Chem. C* **2013**, *117*, 22285-22289.

[17] M. Oehzelt, N. Koch, G. Heimel, *Nat. Commun.* **2014**, *5*, 4174.

[18] J. C. Scott, G. G. Malliaras, *Chem. Phys. Lett.* **1999**, *299*, 115-119.

[19] V. I. Arkhipov, E. V. Emelianova, Y. H. Tak, H. Bässler, *J. Appl. Phys.* **1998**, *84*, 848.

[20] F. So, *Organic Electronics: Materials, Processing, Devices and Applications*, CRC Press, Boca Raton, Florida (USA), **2009**.

[21] C. D. Dimitrakopoulos, P. R. L. Malenfant, *Adv. Mater.* **2002**, *14*, 99-117.

[22] P. M. Langevin, *Ann. Chim. Phys.* **1903**, *28*, 289-383.

[23]   J. Kalinowski, W. Stampor, J. Szmytkowski, D. Virgili, M. Cocchi, V. Fattori, C. Sabatini, *Phys. Rev. B* **2006**, *74*, 085316.

[24]   R. H. Friend, R. W. Gymer, A. B. Holmes, J. H. Burroughes, R. N. Marks, C. Taliani, D. D. C. Bradley, D. A. D. Santos, J. L. Bredas, M. Logdlund, W. R. Salaneck, *Nature* **1999**, *397*, 121-128.

[25]   H. Yersin, A. F. Rausch, R. Czerwieniec, T. Hofbeck, T. Fischer, *Coord. Chem. Rev.* **2011**, *255*, 2622-2652.

[26]   M. Wallesch, D. Volz, D. M. Zink, U. Schepers, M. Nieger, T. Baumann, S. Bräse, *Chem. Eur. J.* **2014**, *20*, 6578-6590.

[27]   J. C. Deaton, S. C. Switalski, D. Y. Kondakov, R. H. Young, T. D. Pawlik, D. J. Giesen, S. B. Harkins, A. J. M. Miller, S. F. Mickenberg, J. C. Peters, *JACS* **2010**, *132*, 9499-9508.

[28]   D. M. Zink, D. Volz, T. Baumann, M. Mydlak, H. Flügge, J. Friedrichs, M. Nieger, S. Bräse, *Chem. Mater.* **2013**, *25*, 4471-4486.

[29]   D. Volz, D. M. Zink, T. Bocksrocker, J. Friedrichs, M. Nieger, T. Baumann, U. Lemmer, S. Bräse, *Chem. Mater.* **2013**, *25*, 3414-3426.

[30]   X. L. Chen, R. M. Yu, Q. K. Zhang, L. J. Zhou, C. Y. Wu, Q. Zhang, C. Z. Lu, *Chem. Mater.* **2013**, *25*, 3910-3920.

[31]   M. Osawa, I. Kawata, R. Ishii, S. Igawa, M. Hashimoto, M. Hoshino, *J. Mater. Chem. C* **2013**, *1*, 4375-4383.

[32]   R. Peng, M. Li, D. Li, *Coord. Chem. Rev.* **2010**, *254*, 1-18.

[33]   R. G. Pearson, *Inorg. Chim. Acta* **1995**, *240*, 93-98.

[34]   D. Saravanabharathi, M. Nethaji, A. G. Samuelson, *Polyhedron* **2002**, *21*, 2793-2800.

[35]   F. H. Jardine, L. Rule, A. G. Vohra, *J. Chem. Soc. A* **1970**, 238-240.

[36]   L. M. Engelhardt, C. Pakawatchai, A. H. White, P. C. Healy, *Dalton Trans.* **1985**, 125-133.

[37]   P. C. Healy, L. M. Engelhardt, V. A. Patrick, A. H. White, *Dalton Trans.* **1985**, 2541-2545.

[38]   J. C. Dyason, P. C. Healy, L. M. Engelhardt, C. Pakawatchai, V. A. Patrick, A. H. White, *Dalton Trans.* **1985**, 839-844.

[39]   L. M. Engelhardt, C. Pakawatchai, A. H. White, P. C. Healy, *Dalton Trans.* **1985**, 117-123.

[40]   J. C. Dyason, P. C. Healy, L. M. Engelhardt, C. Pakawatchai, V. A. Patrick, C. L. Raston, A. H. White, *Dalton Trans.* **1985**, 831-838.

[41]   C. Hirtenlehner, U. Monkowius, *Inorg. Chem. Commun.* **2012**, *15*, 109-112.

[42]   H. Araki, K. Tsuge, Y. Sasaki, S. Ishizaka, N. Kitamura, *Inorg. Chem.* **2007**, *46*, 10032-10034.

[43]   D. Volz, M. Nieger, J. Friedrichs, T. Baumann, S. Bräse, *Inorg. Chem. Commun.* **2013**, *37*, 106-109.

[44]   R. Ahuja, M. Nethaji, A. G. Samuelson, *J. Organomet. Chem.* **2009**, *694*, 1144-1152.

[45]   A. Kaeser, M. Mohankumar, J. Mohanraj, F. Monti, M. Holler, J. J. Cid, O. Moudam, I.
        Nierengarten, L. Karmazin-Brelot, C. Duhayon, B. Delavaux-Nicot, N. Armaroli, J. F.
        Nierengarten, *Inorg. Chem.* **2013**, *52*, 12140-12151.

[46]   C. C. Phifer, D. R. McMillin, *Inorg. Chem.* **1986**, *25*, 1329-1333.

[47]   L. N. Ashbrook, C. M. Elliott, *J. Phys. Chem. C* **2013**, *117*, 3853-3864.

[48]   D. M. Zink, M. Bächle, T. Baumann, M. Nieger, M. Kuhn, C. Wang, W. Klopper, U.
        Monkowius, T. Hofbeck, H. Yersin, S. Bräse, *Inorg. Chem.* **2013**, *52*, 2292-2305.

[49]   D. M. Zink, T. Baumann, J. Friedrichs, M. Nieger, S. Bräse, *Inorg. Chem.* **2013**, *52*,
        13509-13520.

[50]   H. Y. Chao, W. Lu, Y. Li, M. C. Chan, C. M. Che, K. K. Cheung, N. Zhu, *JACS* **2002**, *124*,
        14696-14706.

[51]   Y. G. Ma, C. M. Che, H. Y. Chao, X. M. Zhou, W. H. Chan, J. C. Shen, *Adv. Mater.* **1999**,
        11, 852-857.

[52]   L. Engelhardt, P. Healy, J. Kildea, A. White, *Aust. J. Chem.* **1989**, *42*, 945-947.

[53]   P. Healy, A. Whittaker, J. Kildea, B. Skelton, A. White, *Aust. J. Chem.* **1991**, *44*, 729-736.

[54]   A. Cingolani, Effendy, D. Martini, C. Pettinari, B. W. Skelton, A. H. White, *Inorg. Chim.
        Acta* **2006**, *359*, 2183-2193.

[55]   G. A. Bowmaker, J. C. Dyason, P. C. Healy, L. M. Engelhardt, C. Pakawatchai, A. H.
        White, *Dalton Trans.* **1987**, 1089-1097.

[56]   E. Lastra, M. P. Gamasa, J. Gimeno, M. Lanfranchi, A. Tiripicchio, *Dalton Trans.* **1989**,
        1499-1506.

[57]   Effendy, C. Di Nicola, M. Fianchini, C. Pettinari, B. W. Skelton, N. Somers, A. H. White,
        Inorg. Chim. Acta **2005**, *358*, 763-795.

[58]   K. Tsuge, *Chem. Lett.* **2013**, *42*, 204-208.

[59]   H. Araki, K. Tsuge, Y. Sasaki, S. Ishizaka, N. Kitamura, *Inorg. Chem.* **2005**, *44*, 9667-
        9675.

[60]   D. Volz, Y. Chen, M. Wallesch, R. Liu, C. Fléchon, D. M. Zink, J. Friedrichs, H. Flügge,
        R. Steininger, J. Göttlicher, C. Heske, L. Weinhardt, S. Bräse, F. So, T. Baumann, *Adv.
        Mater.* **2015**, *17*, 2538–2543

[61]   M. J. Leitl, F. R. Küchle, H. A. Mayer, L. Wesemann, H. Yersin, *J. Phys. Chem. A* **2013**,
        117, 11823-11836.

[62]   A. Kobayashi, K. Komatsu, H. Ohara, W. Kamada, Y. Chishina, K. Tsuge, H. C. Chang,
        M. Kato, *Inorg. Chem.* **2013**, *52*, 13188-13198.

[63]   R. W. G. Wyckoff, E. Posnjak, *JACS* **1922**, *44*, 30-36.

[64]   P. Healy, J. Kildea, B. Skelton, A. White, *Aust. J. Chem.* **1989**, *42*, 93-105.

[65]   J. C. Dyason, P. C. Healy, L. M. Engelhardt, C. Pakawatchai, V. A. Patrick, C. L. Raston,
        A. H. White, *Dalton Trans.* **1985**, 831-838.

[66]   P. F. Barron, J. C. Dyason, L. M. Engelhardt, P. C. Healy, A. H. White, *Inorg. Chem.* **1984**, *23*, 3766-3769.

[67]   L. Engelhardt, P. Healy, J. Kildea, A. White, *Aust. J. Chem.* **1989**, *42*, 895-905.

[68]   P. C. Healy, C. Pakawatchai, A. H. White, *Dalton Trans.* **1983**, 1917-1927.

[69]   J. C. Dyason, L. M. Engelhardt, P. C. Healy, C. Pakawatchai, A. H. White, *Inorg. Chem.* **1985**, *24*, 1950-1957.

[70]   P. Healy, B. Skelton, A. Waters, A. White, *Aust. J. Chem.* **1991**, *44*, 1049-1059.

[71]   M. R. Churchill, K. L. Kalra, *Inorg. Chem.* **1974**, *13*, 1065-1071.

[72]   H. Krautscheid, N. Emig, N. Klaassen, P. Seringer, *Dalton Trans.* **1998**, 3071-3078.

[73]   C. Pettinari, C. di Nicola, F. Marchetti, R. Pettinari, B. W. Skelton, N. Somers, A. H. White, W. T. Robinson, M. R. Chierotti, R. Gobetto, C. Nervi, *Eur. J. Inorg. Chem.* **2008**, *2008*, 1974-1984.

[74]   M. Wallesch, D. Volz, C. Fléchon, D. M. Zink, S. Bräse, T. Baumann, *Proc. SPIE 9183* **2014**, 918309.

[75]   T. Banerjee, N. N. Saha, *Acta Cryst. C* **1986**, *42*, 1408-1411.

[76]   D. Benito-Garagorri, W. Lackner-Warton, C. M. Standfest-Hauser, K. Mereiter, K. Kirchner, *Inorg. Chim. Acta* **2010**, *363*, 3674-3679.

[77]   J. Langer, H. Görls, G. Gillies, D. Walther, *Z. anorg. allg. Chem.* **2005**, *631*, 2719-2726.

[78]   S. Owens, Jr., A. Kaisare, G. Gray, *J. Chem. Crystallogr.* **2007**, *37*, 655-661.

[79]   E. H. Wong, F. C. Bradley, L. Prasad, E. J. Gabe, *J. Organomet. Chem.* **1984**, *263*, 167-177.

[80]   F. C. Bradley, E. H. Wong, E. J. Gabe, F. L. Lee, Y. Lepage, *Polyhedron* **1987**, *6*, 1103-1110.

[81]   K. Naktode, R. K. Kottalanka, H. Adimulam, T. K. Panda, *J. Coord. Chem.* **2014**, *67*, 3042-3053.

[82]   E. H. Wong, L. Prasad, E. J. Gabe, F. C. Bradley, *J. Organomet. Chem.* **1982**, *236*, 321-331.

[83]   J. Coetzee, G. R. Eastham, A. M. Slawin, D. J. Cole-Hamilton, *Dalton Trans* **2014**, *43*, 3479-3491.

[84]   C. Zeiher, J. Mohyla, I. P. Lorenz, W. Hiller, *J. Organomet. Chem.* **1985**, *286*, 159-170.

[85]   A. Renz, M. Penney, R. Feazell, K. Klausmeyer, *J. Chem. Crystallogr.* **2012**, *42*, 1129-1137.

[86]   C. S. Kraihanzel, C. M. Bartish, *JACS* **1972**, *94*, 3572-3575.

[87]   G. M. Gray, C. S. Kraihanzel, *J. Organomet. Chem.* **1982**, *238*, 209-222.

[88]   D. Volz, T. Baumann, H. Flügge, M. Mydlak, T. Grab, M. Bächle, C. Barner-Kowollik, S. Bräse, *J. Mater. Chem.* **2012**, *22*, 20786-20790.

[89]   D. Volz, Dissertation, Karlsruhe Institute of Technology (KIT) (Logos Verlag, Berlin), **2014**.

[90]   U. Monkowius, M. Zabel, M. Fleck, H. Yersin, *Z. Naturforsch. B* **2009**, *64*, 1513-1524.

[91]   P. Day, N. Sanders, *J. Chem. Soc. A* **1967**, 1530-1536.

[92]   P. D. Burns, G. N. La Mar, *JACS* **1979**, *101*, 5849-5849.

[93]   R. A. Rader, D. R. McMillin, M. T. Buckner, T. G. Matthews, D. J. Casadonte, R. K. Lengel, S. B. Whittaker, L. M. Darmon, F. E. Lytle, *JACS* **1981**, *103*, 5906-5912.

[94]   I. Abdellah, E. Bernoud, J. F. Lohier, C. Alayrac, L. Toupet, C. Lepetit, A. C. Gaumont, *Chem. Commun.* **2012**, *48*, 4088-4090.

[95]   S. Al-Fayez, L. H. Abdel-Rahman, A. M. Shemsi, Z. S. Seddigi, M. Fettouhi, *J. Chem. Crystallogr.* **2007**, *37*, 517-521.

[96]   A. Mitrofanov, M. Manowong, Y. Rousselin, S. Brandès, R. Guilard, A. Bessmertnykh-Lemeune, P. Chen, K. M. Kadish, N. Goulioukina, I. Beletskaya, *Eur. J. Inorg. Chem.* **2014**, *2014*, 3370-3386.

[97]   R. Starosta, M. Puchalska, J. Cybinska, M. Barys, A. V. Mudring, *Dalton Trans.* **2011**, *40*, 2459-2468.

[98]   X. Liu, J. A. Henderson, T. Sasaki, Y. Kishi, *JACS* **2009**, *131*, 16678-16680.

[99]   D. Volz, M. Nieger, J. Friedrichs, T. Baumann, S. Bräse, *Langmuir* **2013**, *29*, 3034-3044.

[100] J. V. Hanna, M. E. Smith, S. N. Stuart, P. C. Healy, *J. Phys. Chem.* **1992**, *96*, 7560-7567.

[101] A. Olivieri, *JACS* **1992**, *114*, 5758-5763.

[102] D. Volz, M. Wallesch, S. L. Grage, J. Göttlicher, R. Steininger, D. Batchelor, T. Vitova, A. S. Ulrich, C. Heske, L. Weinhardt, T. Baumann, S. Bräse, *Inorg. Chem.* **2014**, *53*, 7838-7847.

[103] L. Monico, K. Janssens, F. Vanmeert, M. Cotte, B. G. Brunetti, G. Van der Snickt, M. Leeuwestein, J. Salvant Plisson, M. Menu, C. Miliani, *Anal. Chem.* **2014**, *86*, 10804-10811.

[104] M. Ralle, S. Lutsenko, N. J. Blackburn, *J. Biol. Chem.* **2003**, *278*, 23163-23170.

[105] I. Jerzykowska, J. Majzlan, M. Michalik, J. Göttlicher, R. Steininger, A. Błachowski, K. Ruebenbauer, *Chem. Erde.* **2014**, *74*, 393-406.

[106] Z. Liu, M. F. Qayyum, C. Wu, M. T. Whited, P. I. Djurovich, K. O. Hodgson, B. Hedman, E. I. Solomon, M. E. Thompson, *JACS* **2011**, *133*, 3700-3703.

[107] Z. Liu, J. Qiu, F. Wei, J. Wang, X. Liu, M. G. Helander, S. Rodney, Z. Wang, Z. Bian, Z. Lu, M. E. Thompson, C. Huang, *Chem. Mater.* **2014**, *26*, 2368-2373.

[108] P. A. M. Dirac, *Proc. R. Soc. A* **1927**, *114*, 243-265.

[109] S. Hüfner, *Photoelectron spectroscopy: Principles and Applications*, Springer-Verlag, Berlin, New York, **1995**.

[110] P. C. Ford, E. Cariati, J. Bourassa, *Chem. Rev.* **1999**, *99*, 3625-3647.

[111]   D. V. O'Connor, W. R. Ware, J. C. Andre, *J. Phys. Chem.* **1979**, *83*, 1333-1343.

[112]   L.-S. Kau, D. J. Spira-Solomon, J. E. Penner-Hahn, K. Hodgson, E. I. Solomon, *JACS* **1987**, *109*, 6433-6442.

[113]   J. E. Hahn, R. A. Scott, K. O. Hodgson, S. Doniach, S. R. Desjardins, E. I. Solomon, *Chem. Phys. Lett.* **1982**, *88*, 595-598.

[114]   I. J. Pickering, G. N. George, C. T. Dameron, B. Kurz, D. R. Winge, I. G. Dance, *JACS* **1993**, *115*, 9498-9505.

[115]   V. J. Catalano, A. L. Moore, J. Shearer, J. Kim, *Inorg. Chem.* **2009**, *48*, 11362-11375.

[116]   E. S. Jeong, J. Park, J. G. Park, D. S. Adipranoto, T. Kamiyama, S. W. Han, *J. Phys.: Condens. Matter.* **2011**, *23*, 175402.

[117]   A. Gaur, B. D. Shrivastava, K. Srivastava, J. Prasad, V. S. Raghuwanshi, *J. Chem. Phys.* **2013**, *139*, 034303.

[118]   J. J. H. Cotelesage, M. J. Pushie, P. Grochulski, I. J. Pickering, G. N. George, *J. Inorg. Biochem.* **2012**, *115*, 127-137.

[119]   T. H. Walther, S. L. Grage, N. Roth, A. S. Ulrich, *JACS* **2010**, *132*, 15945-15956.

[120]   O. Crespo, M. C. Gimeno, A. Laguna, C. Larraz, *Z. Naturforsch. B* **2009**, *64*, 1525-1534.

[121]   J. A. Tang, B. D. Ellis, T. H. Warren, J. V. Hanna, C. L. B. Macdonald, R. W. Schurko, *JACS* **2007**, *129*, 13049-13065.

[122]   R. W. King, T. J. Huttemann, J. G. Verkade, *Chem. Commun.* **1965**, 561a-561a.

[123]   T. J. Penfold, S. Karlsson, G. Capano, F. A. Lima, J. Rittmann, M. Reinhard, M. H. Rittmann-Frank, O. Braem, E. Barano, R. Abela, I. Tavernelli, U. Rothlisberger, C. J. Milne, M. Chergui, *J. Phys. Chem. A* **2013**, *117*, 4591-4601.

[124]   L. Zhang, I. J. Pickering, D. R. Winge, G. N. George, *Chem. Biodiversity* **2008**, *5*, 2042-2049.

[125]   C. Y. Xiang, N. Chopra, J. Wang, C. Brown, S. H. Ho, M. Mathai, F. So, *Org. Electron.* **2014**, *15*, 1702-1706.

[126]   Q. S. Zhang, T. Komino, S. P. Huang, S. Matsunami, K. Goushi, C. Adachi, *Adv. Funct. Mater.* **2012**, *22*, 2327-2336.

[127]   M. Hashimoto, S. Igawa, M. Yashima, I. Kawata, M. Hoshino, M. Osawa, *JACS* **2011**, *133*, 10348-10351.

[128]   H. Uoyama, K. Goushi, K. Shizu, H. Nomura, C. Adachi, *Nature* **2012**, *492*, 234-238.

[129]   G. Liaptsis, D. Hertel, K. Meerholz, *Angew. Chem.* **2013**, *52*, 9563-9567.

[130]   C. Murawski, K. Leo, M. C. Gather, *Adv. Mater.* **2013**, *25*, 6801-6827.

[131]   P. D. Johnson, Y. Ma, *Phys. Rev. B* **1994**, *49*, 5024-5027.

[132]   Y. Ma, *Phys. Rev. B* **1994**, *49*, 5799-5805.

[133]   L. Weinhardt, O. Fuchs, D. Batchelor, M. Bär, M. Blum, J. D. Denlinger, W. Yang, A. Scholl, F. Reinert, E. Umbach, C. Heske, *J. Chem. Phys.* **2011**, *135*, 104705.

[134]   L. Weinhardt, O. Fuchs, A. Fleszar, M. Bär, M. Blum, M. Weigand, J. Denlinger, W. Yang, W. Hanke, E. Umbach, C. Heske, *Phys. Rev. B* **2009**, *79*, 165305.

[135]   A. Benkert, M. Blum, F. Meyer, R. G. Wilks, W. Yang, M. Bär, F. Reinert, C. Heske, L. Weinhardt, *Rev. Sci. Instrum.* **2014**, *85*, 015119.

[136]   F. Meyer, M. Blum, A. Benkert, D. Hauschild, S. Nagarajan, R. G. Wilks, J. Andersson, W. Yang, M. Zharnikov, M. Bär, C. Heske, F. Reinert, L. Weinhardt, *J. Phys. Chem. B* **2014**, *118*, 13142-13150.

[137]   L. Weinhardt, M. Weigand, O. Fuchs, M. Bär, M. Blum, J. Denlinger, W. Yang, E. Umbach, C. Heske, *Phys. Rev. B* **2011**, *84*, 104202.

[138]   L. Weinhardt, A. Benkert, F. Meyer, M. Blum, R. G. Wilks, W. L. Yang, M. Bär, F. Reinert, C. Heske, *J. Chem. Phys.* **2012**, *136*, 144311.

[139]   L. Weinhardt, O. Fuchs, M. Blum, M. Bär, M. Weigand, J. D. Denlinger, Y. Zubavichus, M. Zharnikov, M. Grunze, C. Heske, E. Umbach, *J. Electron Spectrosc. Relat. Phenom.* **2010**, *177*, 206-211.

[140]   J.-P. Charland, J.-L. Roustan, N. Ansari, *Acta Cryst. C* **1989**, *45*, 680-681.

[141]   M. Magnuson, L. Yang, J. H. Guo, C. Såthe, A. Agui, J. Nordgren, Y. Luo, H. Ågren, N. Johansson, W. R. Salaneck, L. E. Horsburgh, A. P. Monkman, *Chem. Phys.* **1998**, *237*, 295-304.

[142]   J. A. Horsley, J. Stöhr, A. P. Hitchcock, D. C. Newbury, A. L. Johnson, F. Sette, *J. Chem. Phys.* **1985**, *83*, 6099.

[143]   G. Vall-llosera, B. Gao, A. Kivimaki, M. Coreno, J. Alvarez Ruiz, M. de Simone, H. Agren, E. Rachlew, *J. Chem. Phys.* **2008**, *128*, 044316.

[144]   C. Kolczewski, R. Püttner, O. Plashkevych, H. Ågren, V. Staemmler, M. Martins, G. Snell, A. S. Schlachter, M. Sant'Anna, G. Kaindl, L. G. M. Pettersson, *J. Chem. Phys.* **2001**, *115*, 6426.

[145]   Å. Hans, A. Reidar, M. Jiri, M. Rolf, *Chem. Phys.* **1984**, *83*, 53-67.

[146]   M. Neeb, A. Kivimäki, B. Kempgens, H. M. Köppe, J. Feldhaus, A. M. Bradshaw, *Phys. Rev. Lett.* **1996**, *76*, 2250-2253.

[147]   M. Neeb, A. Kivimäki, B. Kempgens, H. M. Köppe, K. Maier, A. M. Bradshaw, N. Kosugi, *Chem. Phys. Lett.* **2000**, *320*, 217–221.

[148]   D. M. Zink, D. Volz, L. Bergmann, M. Nieger, S. Bräse, H. Yersin, T. Baumann, *Proc. SPIE 8829* **2013**, 882907.

[149]   S. P. Tandon, J. P. Gupta, *Phys. Status Solidi B* **1970**, *38*, 363-367.

[150]   K. Hattori, T. Mori, H. Okamoto, Y. Hamakawa, *Phys. Rev. Lett.* **1988**, *60*, 825-827.

[151]   R. López, R. Gómez, *J. Sol-Gel Sci. Technol.* **2011**, *61*, 1-7.

[152]   Z. Chen, T. F. Jaramillo, T. G. Deutsch, A. Kleiman-Shwarsctein, A. J. Forman, N. Gaillard, R. Garland, K. Takanabe, C. Heske, M. Sunkara, E. W. McFarland, K. Domen, E. L. Miller, J. A. Turner, H. N. Dinh, *J. Mater. Res.* **2011**, *25*, 3-16.

[153] M.-M. Duvenhage, M. Ntwaeaborwa, H. G. Visser, P. J. Swarts, J. C. Swarts, H. C. Swart, *Opt. Mater.* **2015**, *42*, 193-198.

[154] Z. Z. You, G. J. Hua, *Vacuum* **2009**, *83*, 984-988.

[155] M. Atlan, Y. S. Ocak, S. Pasa, H. Temel, A. Tombak, T. Kilicoğlu, K. Akkilic, M. Aydemir, *Appl. Organomet. Chem.* **2015**, 10.1002/aoc.3371.

[156] T. Phukan, D. Kanjilal, T. D. Goswami, H. L. Das, *Radiat. Meas.* **2003**, *36*, 611-614.

[157] I. Seguy, P. Jolinat, P. Destruel, J. Farenc, R. Mamy, H. Bock, J. Ip, T. P. Nguyen, *J. Appl. Phys.* **2001**, *89*, 5442.

[158] J. Heinze, *Angew. Chem.* **1984**, *96*, 823-840.

[159] S. Admassie, O. Inganäs, W. Mammo, E. Perzon, M. R. Andersson, *Synth. Met.* **2006**, *156*, 614-623.

[160] A. P. Kulkarni, C. J. Tonzola, A. Babel, S. A. Jenekhe, *Chem. Mater.* **2004**, *16*, 4556-4573.

[161] S. M. Abd el Haleem, B. G. Ateya, *J. Electroanal. Chem. Interfacial Electrochem.* **1981**, *117*, 309-319.

[162] A. M. Leiva, L. Rivera, B. Loeb, *Polyhedron* **1991**, *10*, 347-350.

[163] J. Qiu, K. Matyjaszewski, L. Thouin, C. Amatore, *Macromol. Chem. Phys.* **2000**, *201*, 1625-1631.

[164] M. Uda, *Jpn. J. Appl. Phys.* **1985**, *24*, 284-288.

[165] H. Hertz, *Ann. Phys.* **1887**, *267*, 983-1000.

[166] D. Briggs, M. P. Seah, *Practical surface analysis : Auger and x-ray photoelectron spectroscopy*, Wiley, Chichester, **1990**.

[167] A. Jablonski, S. Tougaard, *J. Vac. Sci. Technol. A* **1990**, *8*, 106-116.

[168] H. Bethe, *Ann. Phys.* **1930**, *397*, 325-400.

[169] S. Tanuma, C. J. Powell, D. R. Penn, *Surf. Interface Anal.* **1991**, *17*, 911-926.

[170] S. Tanuma, C. J. Powell, D. R. Penn, *Surf. Interface Anal.* **1994**, *21*, 165-176.

[171] P. van der Heide, *X-ray Photoelectron Spectroscopy*, Wiley, **2011**.

[172] V. Dose, *Appl. Phys.* **1977**, *14*, 117-118.

[173] V. Dose, *Surf. Sci. Rep.* **1985**, *5*, 337-378.

[174] N. V. Smith, *Rep. Prog. Phys.* **1988**, *51*, 1227.

[175] J. Tauc, R. Grigorovici, A. Vancu, *Phys. Status Solidi B* **1966**, *15*, 627-637.

[176] E. O. Kane, *Phys. Rev.* **1962**, *127*, 131-141.

[177] T. Hirooka, M. Kochi, J.-i. Aihara, H. Inokuchi, Y. Harada, *Bull. Chem. Soc. Jpn.* **1969**, *42*, 1481-1486.

[178]   H. Peisert, M. Knupfer, T. Schwieger, J. M. Auerhammer, M. S. Golden, J. Fink, *J. Appl. Phys.* **2002**, *91*, 4872.

[179]   L. Chkoda, C. Heske, M. Sokolowski, E. Umbach, *Appl. Phys. Lett.* **2000**, *77*, 1093.

[180]   D. E. Eastman, *Phys. Rev. B* **1970**, *2*, 1-2.

[181]   H. B. Michaelson, *J. Appl. Phys.* **1977**, *48*, 4729-4733.

[182]   M. Malicki, Z. Guan, S. D. Ha, G. Heimel, S. Barlow, M. Rumi, A. Kahn, S. R. Marder, *Langmuir* **2009**, *25*, 7967-7975.

[183]   D. Marchand, C. Frétigny, M. Laguës, F. Batallan, C. Simon, I. Rosenman, R. Pinchaux, *Phys. Rev. B* **1984**, *30*, 4788-4795.

[184]   F. Maeda, T. Takahashi, H. Ohsawa, S. Suzuki, H. Suematsu, *Phys. Rev. B* **1988**, *37*, 4482-4488.

[185]   S. Suzuki, C. Bower, Y. Watanabe, O. Zhou, *Appl. Phys. Lett.* **2000**, *76*, 4007-4009.

[186]   L. Stolberg, J. Richer, J. Lipkowski, D. E. Irish, *J. Electroanal. Chem. Interfacial Electrochem.* **1986**, *207*, 213-234.

[187]   L. Stolberg, J. Lipkowski, D. E. Irish, *J. Electroanal. Chem. Interfacial Electrochem.* **1987**, *238*, 333-353.

[188]   A. Bilić, J. R. Reimers, N. S. Hush, *J. Phys. Chem. B* **2002**, *106*, 6740-6747.

[189]   S. Krause, M. B. Casu, A. Schöll, E. Umbach, *New J. Phys.* **2008**, *10*, 085001.

[190]   W. C. Still, M. Kahn, A. Mitra, *J. Org. Chem.* **1978**, *43*, 2923-2925.

[191]   G. Sheldrick, *Acta Cryst. A* **2008**, *64*, 112-122.

[192]   G. Sheldrick, *Acta Cryst. C* **2015**, *71*, 3-8.

[193]   N. C. Greenham, I. D. W. Samuel, G. R. Hayes, R. T. Phillips, Y. A. R. R. Kessener, S. C. Moratti, A. B. Holmes, R. H. Friend, *Chem. Phys. Lett.* **1995**, *241*, 89-96.

[194]   G. Metz, X. L. Wu, S. O. Smith, *J. Magn. Reson. A* **1994**, *110*, 219-227.

[195]   N. M. Szeverenyi, M. J. Sullivan, G. E. Maciel, *J. Magn. Reson.* **1982**, *47*, 462-475.

[196]   H. G. O. Becker, *Organikum*, Deutscher Verlag der Wissenschaften, Berlin, **1990**.

[197]   C.-H. Lin, Y. Chi, M.-W. Chung, Y.-J. Chen, K.-W. Wang, G.-H. Lee, P.-T. Chou, W.-Y. Hung, H.-C. Chiu, *Dalton Trans.* **2011**, *40*, 1132-1143.

[198]   M. Newville, *J. Synchrotron Radiat.* **2001**, *8*, 96-100.

[199]   B. Ravel, M. Newville, *J. Synchrotron Radiat.* **2005**, *12*, 537-541.

[200]   M. S. Moreno, K. Jorissen, J. J. Rehr, *Micron* **2007**, *38*, 1-11.

[201]   A. L. Ankudinov, A. I. Nesvizhskii, J. J. Rehr, *Phys. Rev. B* **2003**, *67*.

[202]   G. Meitzner, G. H. Via, F. W. Lytle, J. H. Sinfelt, *J. Chem. Phys.* **1985**, *83*, 353-360.

[203]   L. Banci, I. Bertini, F. Cantini, S. Ciofi-Baffoni, L. Gonnelli, S. Mangani, *J. Biol. Chem.* **2004**, *279*, 34833-34839.

[204]   S. W. Han, E. A. Stern, D. Haskel, A. R. Moodenbaugh, *Phys. Rev. B* **2002**, *66*, 094101.

[205]   M. Blum, L. Weinhardt, O. Fuchs, M. Bär, Y. Zhang, M. Weigand, S. Krause, S. Pookpanratana, T. Hofmann, W. Yang, J. D. Denlinger, E. Umbach, C. Heske, *Rev. Sci. Instrum.* **2009**, *80*, 123102.

[206]   R. N. S. Sodhi, C. E. Brion, *J. Electron Spectrosc. Relat. Phenom.* **1984**, *34*, 363-372.

8 Abbreviations

approx. – approximately

bn – billion, 10^9

CuAAC – Cu(I)-catalyzed azide alkyne cycloaddition

CV – cyclic voltammetry

dmp – 2,9-Dimethyl-1,10-phenanthroline

DOS – density of states

EA – electron affinity

EBL – electron blocking layer

EI – electron ionization

EL – emissive layer

EQE – external quantum efficiency

ETL – electron transport layer

EXAFS – extended x-ray absorption fine structure

F – Fermi level

FAB – fast atom bombardment

HBL – hole blocking layer

HIL – hole injection layer

HOMO – highest occupied molecular orbital

HPLC – high-performance liquid chromatography

HRMS – high-resolution mass spectrometry

HTL – hole transport layer

IEELS – inner shell electron energy loss spectroscopy

IP – ionization potential

IPES – inverse photoemission spectroscopy

IQE – internal quantum efficiency

ITO – indium tin oxide

IUPAC – International Union of Pure and Applied Chemistry

LCD – liquid crystal display

LDA – lithium diisopropylamide

LED – light emitting diode

LUMO – lowest unoccupied molecular orbital

MS – mass spectrometry

NHetPHOS – N heteroaromatic diphenyl-phosphine ligand

NMR – Nuclear magnetic resonance spectroscopy

NMR CP/MAS – nuclear magnetic resonance (NMR) in cross polarization (CP) using magic angle spinning (MAS)

OLED – organic light emitting diode

opt – optical

PES – photoelectron spectroscopy

PESA – photoelectron spectroscopy in air

phen – 1,10-Phenanthroline

PLQY – photoluminescence quantum efficiency

POP – tetraphenyldiphosphoxane

POQ – 8-((diphenylphosphanyl)oxy)-quinoline

PPO – tetraphenyldiphosphine monoxide

RIXS – resonant inelastic soft x-ray scattering

SCE – saturated calomel electrode

TADF – thermally activated delayed fluorescence

UHV – ultra high vacuum

UPS – ultraviolet photoelectrons spectroscopy

UV – ultraviolet

vac – vacuum

w.r.t. – with respect to

XANES – x-ray absorption near edge structure

XAS – x-ray absorption

XRD – single crystal x-ray diffraction

9   Appendix

9.1   Scientific career

2012–2015   **Doctoral research studies** at the **Karlsruhe Institute of Technology (KIT)**, Germany, Institute of Organic Chemistry and Institute for Photon Science and Synchrotron Radiation, Supervisor: Prof. Dr. Stefan Bräse, Co-Supervisor Dr. Lothar Weinhardt

2014   **Six-month research project** at the **University of Nevada, Las Vegas (UNLV)**, USA in collaboration with Prof. Dr. Clemens Heske

2013–2015   Doctoral scholarship of the **Deutsch Telekom Stiftung**

2012   Funding through the **Feasibility Studies of Young Scientists** from KHYS

2009–2012   KIT, Faculty of Chemistry and Biosciences, **Diploma in chemistry, final grade: 1.1**

2007–2009   KIT, Faculty of Chemistry and Biosciences, **Intermediate diploma in chemistry, final grade: 1.0**

2004–2007   Albertus Magnus secondary school Ettlingen, General qualification for university entrance, **final grade: 1.0**

9.2 Publications, conference contributions

9.2.1 Publications (peer reviewed)

D. Volz, Y. Chen, **M. Wallesch**, R. Liu, C. Fléchon, D. M. Zink, J. Friedrichs, H. Flügge, R. Steininger, J. Göttlicher, C. Heske, L. Weinhardt, S. Bräse, F. So, T. Baumann, Thomas, Bridging the Efficiency Gap: Fully Bridged Dinuclear Cu(I)-Complexes for Singlet Harvesting in High-Efficiency OLEDs, *Adv. Mater.* **2015**, *27*, 1521-4095.

D. Volz, **M. Wallesch**, C. Fléchon, M. Danz, A. Verma, J. M. Navarro, D. M. Zink, S. Bräse, T. Baumann, From iridium and platinum to copper and carbon: new avenues for more sustainability in organic light-emitting diodes, *Green Chem.* **2015**, *17*, 1988-2011.

M. Wallesch, D. Volz, D. M. Zink, U. Schepers, M. Nieger, T. Baumann, S. Bräse, Bright Coppertunities: Multinuclear CuI Complexes with N-P Ligands and Their Applications, *Chem. Eur. J.* **2014**, *22*, 6578-6590.

D. Volz, **M. Wallesch**, S. L. Grage, J. Göttlicher, R. Steininger, D. Batchelor, T. Vitova, A. S. Ulrich, C. Heske, L. Weinhardt, T. Baumann, S. Bräse, Labile or stable: Can homoleptic and heteroleptic PyrPHOS-copper complexes be processed from solution?, *Inorg. Chem.* **2014**, *53*, 7838-7847.

D. M. Zink, L. Bergmann, D. Ambrosek, **M. Wallesch**, D. Volz, M. Mydlak, Singlet harvesting copper-based emitters: a modular approach towards next-generation OLED technology, *Transl. Mater. Res.* **2014**, *1*, 015003.

K. S. Masters, **M. Wallesch**, S. Bräse, ortho-Bromo(propa-1,2-dien-1-yl)arenes: Substrates for Domino Reactions, *J. Org. Chem.* **2011**, *76*, 9060-9067.

9.2.2 Publications (other)

D. Volz, **M. Wallesch**, S. Bräse, T. Baumann, Late bloomers: copper complexes in organic LEDs, *SPIE Newsroom* **2014**, doi: 10.1117/2.1201409.005617.

9.2.3 Conference contributions

Posters

M. Wallesch, D. Volz, D. Batchelor, T. Vitova, R. Steininger, J. Göttlicher, S. Bräse, C. Heske, L. Weinhardt, Cu K XANES and EXAFS spectroscopy on Cu(I)-emitters, XAFS16 25.08.2015, Karlsruhe (Germany).

M. Wallesch, D. Volz, S. Bräse, L. Weinhardt, C. Heske, T. Baumann, Printed organic light-emitting devices with copper(I)-emitters: towards more sustainability in optoelectronics, JCF 26.03.2015, Münster (Germany).

M. Wallesch, L. Weinhardt, D. Kreikemeyer-Lorenzo, A. Benkert, F. Meyer, D. Volz, M. Blum, M. Bär, R. G. Wilks, W. Yang, S. Bräse, C. Heske, Investigation of luminescent copper(I) complexes via resonant inelastic soft x-ray scattering, ANKA/KNMF User Meeting 26.09.2013, Bruchsal (Germany).

D. Volz, A. Jacob, **M. Wallesch**, C. Fléchon, J. M. Navarro, D. M. Zink, J. Friedrichs, D. Ambrosek, G. Liaptsis, S. Bräse, T. Baumann, Efficient dinuclear copper(I)-complexes for solution- and vacuum-deposited organic light-emitting devices (OLEDs), Faraday Discussion 08.08.2014 in Galsgow (United Kingdom).

Proceedings

M. Wallesch, S. Bräse, T. Baumann, D. Volz, X-ray absorption spectroscopy: towards more reliable models in material sciences, *Proc. SPIE 9566* **2015**, 956609, San Diego (USA).

H. Flügge, A. Rohr, S. Döring, C. Fléchon, **M. Wallesch**, D. Zink, J. Seeser, J. Leganés, T. Sauer, T. Rabe, W. Kowalsky, T. Baumann, D. Volz, Reduced concentration quenching in a TADF-type copper(I)-emitter, *Proc. SPIE 9566* **2015**, 95661, San Diego (USA).

M. Wallesch, D. Volz, C. Fléchon, D. M. Zink, S. Bräse, T. Baumann, Bright coppertunities: efficient OLED devices with copper(I)iodide-NHetPHOS-emitters, *Proc. SPIE 9183* **2014**, 918309, San Diego (USA).

9.3   Acknowledgement

I want to thank my supervisors Prof. Dr. Stefan Bräse, Dr. Lothar Weinhardt, and Prof. Dr. Clemens Heske, who offered me the opportunity to work on a fascinating topic during my doctoral research studies. Thank you for all the time you spent on discussions of my results.

I owe gratitude to the Deutsche Telekom Stiftung, which supported me with a scholarship during my doctoral research studies. The Research Travel Grant from the KHYS enabled my six-month research project at the UNLV and the funding through the Feasibility Studies of Young Scientists from KHYS supported the synthetic part of my doctoral research studies.

Furthermore, I thank my collaboration partners who contributed to this work in manifold ways: The Heske group in Karlsruhe, Würzburg, and Las Vegas helped me with the PES and soft x-ray spectroscopic experiments and data analysis. Dr. Ralph Steininger, Dr. Jörg Göttlicher, and Dr. habil. Anton Plech from ANKA spent a great effort and amount of time on the hard x-ray spectroscopy experiments. Dr. Tonya Vitova from INE and Dr. David Batchelor from ANKA were always willing to discuss my results and helped me with the analysis of the x-ray spectroscopic data. CYNORA provided ligands and Cu(I) complexes as well as help and the experimental setup for the cyclic voltammetry and photophysical measurements. Dr. Martin Nieger carried out the single crystal x-ray diffraction experiments and analysis. Tanja Pürckhauer from the LTI performed the PESA measurements.

Daniel – thank you for your support throughout the past years.